A History of Life
in 100 Fossils

Paul D. Taylor & Aaron O'Dea

Published by the Natural Hstory Museum, London

This book is dedicated to our own treasured specimens:
Mila and Lorenzo O'Dea and Emma and James Taylor.

Paul D. Taylor worked at the Natural History Museum, London for 39 years, heading the Invertebrate and Plants division between 1990 and 2003. His research focuses on fossil and living bryozoans, with subsidiary interests in evolution, palaeoecology and fossil folklore. He is the author of the book *Fossil Invertebrates*.

Aaron O'Dea is a palaeobiologist at the Smithsonian Tropical Research Institute in Panama. He uses the fossil record to understand the drivers of evolution and to reconstruct what Caribbean reefs were like before humans. Aaron has a broad passion for exploring, understanding and communicating the history of life in the tropics.

First published in hardback by the Natural History Museum
Cromwell Road. London SW7 5BD
© The Trustees of the Natural History Museum, London 2014

This edition published by the Natural History Museum, London, 2018 and 2024

ISBN 978 0 565 09549 9

10 9 8 7 6 5 4 3

Special photography by Harry Taylor, Natural History Museum, London
Designed by Mercer Design, London
Reproduction by Saxon Digital Services, UK
Printed by Toppan Leefung Printing Limited

CONTENTS

Introduction

THE HISTORY OF LIFE IS WRITTEN in the rocks, or more precisely in the fossils contained within the rocks. Fossils are the remains of ancient animals and plants, as well as fungi and microbes, preserved by natural processes of burial and entombment. Without the existence of fossils it is doubtful whether Charles Darwin would ever have devised his theory of evolution by natural selection. The fossils collected by Darwin in South America during his formative years as a naturalist aboard HMS *Beagle* convinced him that the Earth was once inhabited by animals and plants unlike any that are living today. Life has changed through time, with old species becoming extinct and new ones evolving from the old. Over 99% of all species that have ever lived are now extinct. Change is a necessary part of evolution. However, it is not simply a matter of new species replacing old ones, as Baron Cuvier and other prominent naturalists at the same time as Darwin believed. For evolution to occur there must also be the modification of existing species into new species.

It is always worth emphasizing that the fossil record, although glorious in what it does preserve, is a very incomplete register. Only a small proportion of species that ever existed have been fossilized and, among those species that have been discovered as fossils, only the tiniest fraction of the individuals that lived will ever be dug up as fossils. Furthermore, soft body parts such as muscles, nerves and guts are in most cases lost during fossilization, and DNA also disappears with time. Nevertheless, when scrutinized with great care the fossil record allows us to detect many of the landmark events in the history of life. Because all of life on Earth shares a common ancestor, the fossil record is a register of just one 'go' at evolution on one planet. Nonetheless, strikingly similar events are repeated across time, and it is these that uncover fundamental evolutionary principles, such as convergent evolution when formerly dissimilar organisms evolve similar shapes as a result of adapting to the same mode of life.

Fossils come in two main varieties: body fossils and trace fossils. Body fossils are the preserved shells, bones, teeth, leaves etc. of the once living organisms. Trace fossils are indications of the activities of organisms, such as footprints, burrows and even excrement. Both types give palaeontologists a great deal of information about which species lived when and where, and how they lived, through almost 4.6 billion years of geological time since the Earth was first formed. While trace fossils can record the behaviours of long dead animals in detail, the exact identity of the tracemaker is seldom known, except in very rare instances where a body fossil is present too.

Inspired by the immense collections in our respective museums, we have selected fossils that highlight milestones in the history of life, from its origins to the emergence of modern humans. Examples of the first and last fossils of their kinds are particularly critical in allowing us to reconstruct the narrative history of life on Earth: when did particular types of animals and plants first evolve, when did those that are now extinct suffer their demise, and can we say anything about why they evolved and became extinct? In part guided by the availability of suitable specimens, we also made selections based on our own personal preferences; other palaeontologists would no doubt have selected an entirely different 100 fossils. We have chosen fossils from across the entire spectrum of life, from animals to plants, microbes to dinosaurs, creatures of the land to inhabitants of the oceans.

The history of life has been enacted in a world of constant change over geological time, with landmasses migrating across the surface of the globe, oceans appearing and disappearing, and sea-levels rising to flood low-lying land and then falling back to expose it once more. In addition, there were more rapid changes. The eruption of volcanoes, often lethal for life on a local scale, has occasionally had a global impact, changing the composition and clarity of the atmosphere and affecting the climate of the entire planet. For whatever reason, global climates have fluctuated wildly through the course of geological time. Polar ice-caps have come and gone, with tropical environments sometimes stretching into much higher latitudes than they do today. And, after much early speculation, scientists are now beginning to discover compelling evidence for extraterrestrial influences on our planet and its life, most notably through the environmental devastation and destruction of ecosystems wrought by the collision of asteroids with the Earth.

Other fossils illustrate evolutionary themes: speciation, evolutionary radiation, invasion, sexual selection, gigantism and dwarfism. Despite important modern advances in genetics and molecular biology, much of what we know about how life on Earth has evolved still comes from fossil evidence. Fossils are witnesses to the past, albeit imperfect witnesses demanding close scrutiny if the story they tell is to be understood properly. For example, the theory of punctuated equilibrium – that many evolutionary lineages remained static for millions of years before changing very rapidly – was devised through detailed studies of fossil lineages. It could not have been found or anticipated by studying living animals and plants alone.

The 100 fossils are arranged roughly in chronological order, from the oldest to the youngest. This has obvious advantages in allowing the history of life to unfold through the pages of the book. The book has been subdivided into four sections representing major divisions of geological time (see p. 226 for the geological timescale): Precambrian (4,600–539 million years), Palaeozoic (539–252 million years), Mesozoic (252–66 million years) and Cenozoic (66 million years to the present-day). These divisions have been recognized by scientists for two centuries and form natural chapters in the story of life's evolution.

PRECAMBRIAN

THE EARTH CONDENSED from a cloud of interstellar dust some 4.6 billion years ago. This point marked the beginning of an enormous eon of geological time called the Precambrian. A time traveller arriving on the Precambrian Earth could be forgiven for believing that he or she had alighted on an alien planet: low levels of oxygen in the air would make breathing impossible, and neither animals nor plants would be in evidence. Seen from space, none of the familiar continents of today existed, the shapes and arrangements of the landmasses being entirely different.

Not surprisingly, given that it lasted 4 billion years, eight times the duration of the succeeding Phanerozoic eon, the Precambrian witnessed profound changes in the Earth's environments. For about the first 500 million years the Earth was in a state of constant bombardment by meteorites, making it extremely inhospitable to life. Eventually, however, the meteor storms abated and our planet acquired a more stable surface. Liquid water and carbon dioxide were both present, but oxygen levels were very low. It was in this setting that life originated, possibly as much as 3.9 billion years ago, although the oldest possible fossils are considerably younger, a mere 3.46 billion years old. These tiny fossils resemble modern cyanobacteria and represent just one of the many kinds of microbial organisms that flourished on the Precambrian Earth. Communities of microbes, including bacteria in addition to cyanobacteria, sometimes combined to trap sediment and produce mound-like layered structures called stromatolites. Only then did they form fossils large enough to be easily noticed. Stromatolites are very much a signature of the Precambrian. They peaked in abundance about 1.25 billion years ago, but can still be found in special places today, such as Shark Bay in Western Australia.

Bacteria and cyanobacteria make energy to live in various ways. One of these is by photosynthesis, a process that releases oxygen into the atmosphere. The oxygen produced by Precambrian organisms was at first soaked up by iron dissolved in the oceans. Oxidized iron in the form of the minerals haematite and magnetite sank to the seabed, resulting in vast banded iron formations. Many of these commercially important deposits were formed during the so-called 'great oxygenation event' that occurred 2.4 billion years ago. Eventually, however, iron to soak up the oxygen became largely exhausted and oxygen was able to accumulate in the oceans and in the atmosphere. Increasing oxygen levels paved the way for more advanced organisms to evolve, including fungi, plants and animals with large, multicellular bodies. Stromatolites fell into decline, victims of grazing animals seeking food.

The exact time when complex animals appeared on the Earth is still a matter of scientific debate. Calibration of the differences between molecular sequences in modern animals has suggested to some biologists that fossils of large animals ought to be present in rocks as much as 1 billion years old. However, as yet the oldest known fossils fitting the bill come from the Ediacaran period (635–539 million years), which marks the very end of the Precambrian. Ediacaran fossils include both miraculously preserved embryos and some peculiar pleated fossils of larger size.

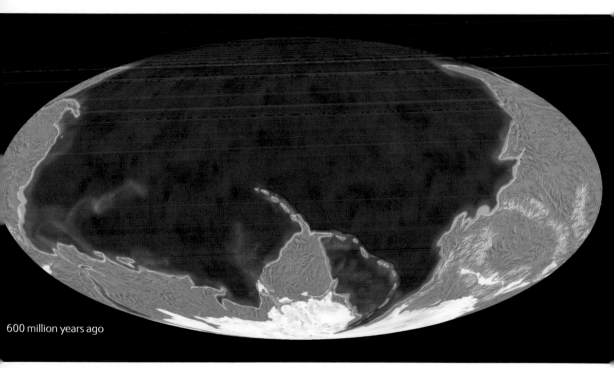

600 million years ago

10 μm

World's oldest fossils?
Apex Chert

THE ORIGIN OF LIFE from a non-living precursor is a question that has taxed the minds of countless scientists, and we are still far from knowing the answer. Here, fossils do not help a great deal. However, they do show us that all living organisms older than 600 million years were small and simple in structure, most resembling modern bacteria, and that these organisms have inhabited the Earth for more than two-thirds of its 4.6-billion-year history.

In 1993 Professor J William Schopf (1941–), a distinguished scholar of ancient life, published a scientific paper describing what might be the world's oldest fossils. Schopf's specimens were collected in Western Australia, from a glassy rock called the Apex Chert, and are estimated to be about 3,465 million years old on the basis of radiometric dates obtained from associated volcanic rocks. They are minute filaments, interpreted by Schopf as probable cyanobacteria. As they are less than one-hundredth of a millimetre across, they cannot be seen on the surface of the rock and only become visible using a microscope after grinding down pieces of chert into wafer-thin slices through which light will pass. These thin sections reveal the presence of abundant sinuous tubular filaments, dark brown or black and often subdivided by cross walls along their length. Schopf was able to distinguish eleven different types of filament, each of which he regarded as a distinct species.

The Apex Chert filaments have become a focus of heated debate among scientists, with some supporting Schopf's contention that they are true fossils but others favouring an inorganic origin. This controversy highlights the broader issue of how biological fossils can be distinguished from non-biological, inorganic structures. Complex fossils offer no difficulty – how, for example, could the intricate skeleton of a trilobite be anything other than the fossil of a once living organism? However, simple structures like the Apex Chert filaments are a different matter altogether. Evidence supporting their interpretation as fossils comes from the close similarity in appearance to some modern cyanobacteria, and also from their chemical composition, which Schopf determined to be kerogen, a type of carbon usually formed from the breakdown of the organic compounds found in living organisms. Critics have pointed out that some of the filaments have a branched structure unlike that of modern cyanobacteria, and have claimed that the 'fossils' are instead hairline fractures in the rock that became filled by inorganic minerals. Although the jury is still out for the Apex Chert filaments, and may be for a very long time to come, other chemical evidence is consistent with the existence of life on Earth 3.5 billion years ago.

Great Oxygenation
stromatolites

STROMATOLITES ARE ARGUABLY one of the most important types of organic structure ever to have existed in the history of life on Earth. Not a single organism, stromatolites are communities formed by the trapping and binding of sediment by many generations of cyanobacteria and other microbes, which constructed distinctive layered columns in rock-like formations.

One of the earliest forms of life on Earth and still alive today, stromatolites have a fossil record dating back an incredible 3.5 billion years, and they may be even older than that. They dominated the world's shallow oceans for a staggering 3 billion years, reproducing asexually by cloning and building thick sedimentary deposits, creating a stromatolite nirvana. Seemingly unstoppable, they used water with carbon dioxide and sunlight to make sugars to fuel their growth – essentially the same photosynthesis used by modern plants – and in the process produced oxygen.

At first, this oxygen combined with the iron that was dissolved in the world's oceans and formed insoluble iron layers, which precipitated out onto the sea floor. These bands of iron form the vast majority of the world's iron ores. Only when most of the dissolved iron in the oceans was consumed could free oxygen levels start to build in the atmosphere, leading to the 'Great Oxygenation Event' (GOE).

The GOE was the breakthrough needed for the evolution of complex oxygen-breathing and sexually reproducing life forms that soon dominated life on Earth. One theory suggests that the anaerobic bacteria in stromatolites could not protect themselves against the grazing of the more complex, oxygen-breathing creatures, and their supremacy came to an end. Indeed, stromatolites today are relegated to extremely salty lagoons, where grazing animals cannot live. It appears that the stromatolites helped build the oxygen-rich world that supports the vast diversity of life we know today. However, in doing so they caused their own downfall – but not extinction – forced to live on the margins of existence.

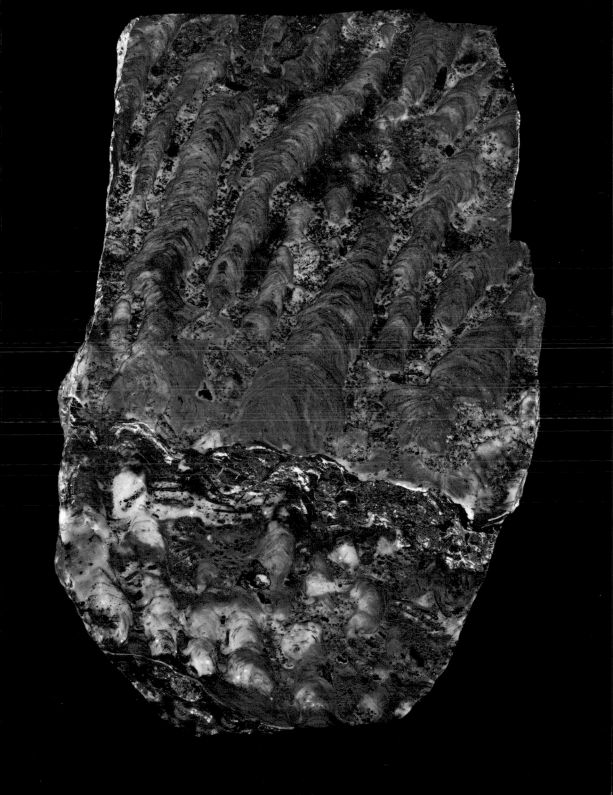

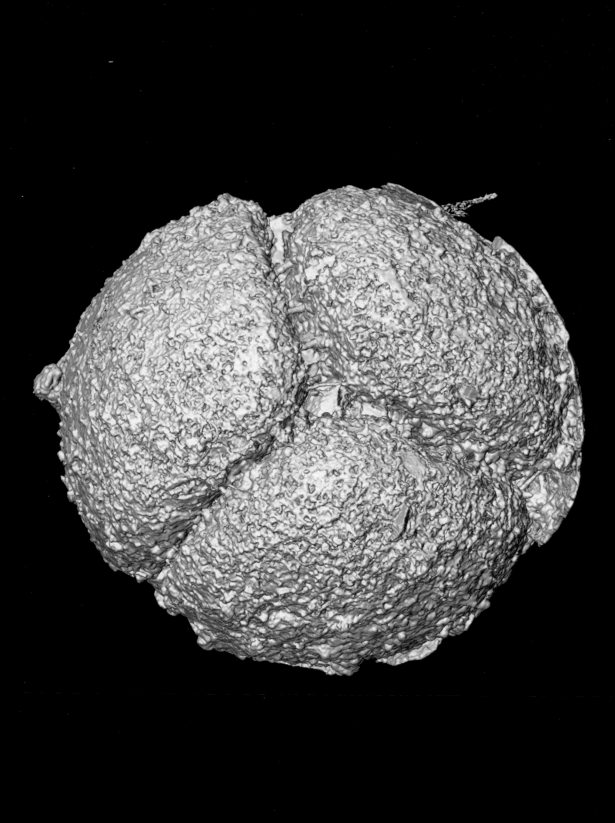

Embryonic enigmas
Doushantuo fossils

THE FOSSIL RECORD DOES a poor service to those animals without a hard skeleton. Only under exceptional circumstances do soft-bodied animals become preserved, and the sites in which such fossils can be found are called fossil Lagerstätten. The 635–551-million-year-old Doushantuo Formation in China is one of the world's most remarkable Lagerstätte. Soft-bodied fossils are preserved in such minute detail that, using high-powered scanning techniques, even individual cells can be identified. The reason for this extraordinary preservation is thanks to a process called phosphatization: the replacement of dying cells by salts, which are then resistant to decay and destruction. The resulting fossils help palaeontologists fill in the blanks when reconstructing the origins of animal life.

Fossilized skeletons of complex, multicellular animals appeared suddenly in the Cambrian Explosion, almost 540 million years ago, but this apparent surge in animal diversity may represent the adoption of hard skeletons rather than a diversification of life itself. Analysis of DNA (the blueprint shared across all known life) backs this up, pointing to the evolution of animals from a single-celled organism with a whip-like structure used for propulsion some 780 million years ago. There is evidently a substantial lag between the origin of animal life and the unequivocal appearance of animals in the fossil record, and that is why the Doushantuo fossils are so important.

In the Doushantuo Formation one can find beautifully preserved fossils that resemble, astonishingly, the embryos of modern animals. They are shaped by adjoining cells, suggesting a flexible membrane rather than the rigid cell walls of algae or fungi. The number of cells is always to the power of two (1, 2, 4, 8, etc.) characteristic of embryo cleavage, and they are larger than other types of cells. These very old soft-bodied fossils are open to a variety of different interpretations, and the idea that they represent the oldest animal fossils has been challenged. As the debate rumbled on, a group of scientists blasted some of the fossils in a high-powered particle accelerator called a synchrotron. The images produced, like the one opposite, are of unprecedented resolution, revealing not only the structure of the multiplying cells, but also structures that look like nuclei inside the cells. If so, their location is incompatible with the idea that they are embryos. So if not bacteria, algae, fungi or animal embryos, what are these remarkable fossils? Some suggest they are a form of life more akin to a parasite that multiplies its cells within a cyst that then explodes them as 'spores' to propagate the next generations.

Precambrian paradox
Dickinsonia

IN 1946 AUSTRALIAN GEOLOGIST Reg Sprigg (1919–1994) was eating his lunch in the Ediacara Hills of South Australia when he made one of the most momentous fossil discoveries of all time. Sprigg found some peculiar, jellyfish-like impressions in rocks about 550 million years old. Predating the Cambrian Explosion (p.18), these Ediacaran fossils still perplex palaeontologists. Similar fossils had been found previously in Newfoundland and Namibia, but these attracted little attention because of doubts about their age or whether they really were fossils. Indeed, it was only after the discovery in Charnwood Forest, Leicestershire, England of another Ediacaran-type fossil, *Charnia*, in rocks firmly dated as Late Precambrian that the Ediacaran fossils took centre stage.

But what is so peculiar about Ediacaran fossils? Firstly, they lack any trace of hard skeletons and are therefore unlike the shelled animals found in slightly younger rocks after the Cambrian Explosion. More importantly, the Ediacaran fossils consistently defy close comparison with any organisms living today. Some of them vaguely resemble flattened jellyfish but lack the tentacles to corroborate this idea. Others, like *Charnia*, are leaf-shaped, leading to an early belief that they were relatives of modern sea pens. However, this idea is untenable because the solid fronds of *Charnia* are unlike the polyp-lined branches of sea pens. Added to these are some far stranger Ediacaran fossils. *Dickinsonia* looks like a disc-shaped worm with a segmented body. Specimens of *Dickinsonia* range in size from a few centimetres to more than a metre (3 feet 3 inches) without changing much in shape, which is itself unusual. Another disc-shaped Ediacaran fossil, *Tribrachidium,* has a three-lobed central structure reminiscent of the three-legged triskell, the emblem of the Isle of Man.

A consensus has yet to be reached about the identity of these bizarre fossils. Are they animals, plants or, as one prominent palaeontologist has suggested, a sort of lichen (a symbiotic association between a fungus and an alga)? Some scientists believe they represent primitive ancestors of animal phyla still living today, whereas others believe them to belong to a distinct group of pleated organisms – the Vendobionta – long since extinct. There is also debate about how they lived. Did they obtain energy through photosynthesis, feed on plankton like many modern animals living in the sea, or subsist by absorbing dissolved organic compounds directly from the seawater? Another school of thought contends that at least some of the Ediacaran fossils were animals that grazed on microbial mats (complex filamentous webs of cyanobacteria and other microorganisms). This hypothesis has added appeal in providing an explanation for the unusual preservation of these soft-bodied organisms. Sticky microbial mats, evidence for which is found in the presence of wrinkled, 'elephant-skin' textures in the sediment, could have facilitated the preservation of imprints of the corpses of animals that once fed on them. Whatever their mode of living, the eventual evolution of more advanced animals capable of churning up the microbial mats in the Cambrian put paid to the unique Ediacaran life forms unearthed during Reg Sprigg's lunch break.

PALAEOZOIC ERA

FROM THE CAMBRIAN EXPLOSION of life, to the great extinction event at the end of the Permian period almost 300 million years later, the Palaeozoic was an era of unprecedented change. Major transformations in the biosphere occurred in parallel with the geological dynamism, the Earth's continents drifting, separating and colliding, driven along by slow but sustained movements of magma deep beneath the surface. The continents of Laurentia and Baltica formed and were later fragmented, while a massive southern continent called Gondwana persisted until the end of the Palaeozoic when much of the Earth's land became assembled into one enormous continent called Pangaea. Early scientific evidence for continental drift came from the distributions of particular types of fossils that appeared anomalous when plotted onto a modern globe but made sense if the continents had once been configured differently.

Climates varied through the Palaeozoic. For much of the era, polar icecaps were absent and the Earth was typically warmer than it is today. However, two main episodes of cooling and glaciation occurred, one at the end of the Ordovician period and the second in the Late Carboniferous and Early Permian. The first of these glaciations has been suggested as the main cause in a mass extinction event that affected marine animals in the latest Ordovician. The cause of other mass extinctions during the Late Devonian is less apparent but these Devonian events restructured the ecology of marine life, though not nearly to such a great extent as the end-Permian mass extinction at the close of the Palaeozoic era.

The Palaeozoic saw the first appearance in the fossil record of most of the phyla of animals we know today, for example, arthropods and echinoderms, as well as the vertebrates. Underwater reefs were constructed by a variety of animals inhabiting the seabed, including corals and sponges. The erroneously named sea-lilies, which are not plants but echinoderms related to starfish, at times carpeted the Palaeozoic seabed with forest-like growths, harbouring symbiotic snails in an evolutionary association that lasted for a remarkable 200 million years. Within these reefs and stony sea-lily forests, as well as in other marine environments, the struggle for life became ever more intense – predators evolved more effective ways of capturing, subduing and consuming their prey, while the victims countered with defences such as spines to ward-off their assailants. Early fishes in the Palaeozoic lacked jaws but eventually fishes with jaws and teeth appeared, becoming the top predators in the sea.

The land surface witnessed its first sizeable inhabitants adding to the microbes that had been around for far longer. Vascular plants – with xylem to conduct water to the leaves and phloem to take organic nutrients in the opposite direction – colonized the land for the first time, soon evolving into trees that formed the first forests. Vertebrates and arthropods, in the guise of amphibians and insects respectively, quickly followed plants onto the land, taking advantage of the new food resources available in this no longer barren environment. Patterns of weathering and erosion became transformed under the influence of plants, which on the one hand provided a protective blanket but on the other promoted the breakdown of rocks and the formation of soils.

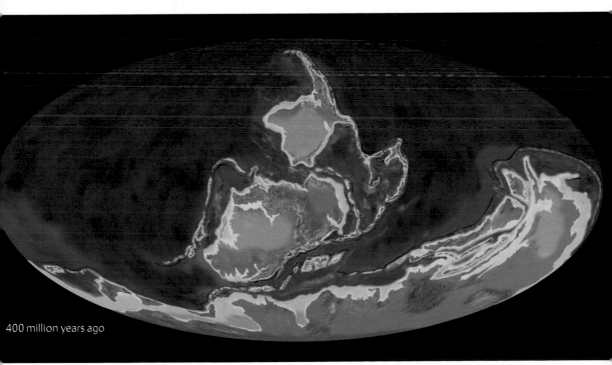

400 million years ago

Cambrian Explosion
Anomalocaris

THE GREATEST BLOSSOMING of life recorded by fossils occurred during the Cambrian period, between 539 and 485 million years ago. The Cambrian Explosion, as it has come to be known, saw the first appearance of fossils belonging to most of the animal phyla that are still around today: molluscs, arthropods, echinoderms, chordates and many others. Our knowledge of the diversification of life in Cambrian seas has been greatly enhanced by fossils from two exceptional deposits – the Burgess Shale of British Columbia, Canada and the slightly older Maotianshan Shales of Chengjiang County, Yunnan, China. These sites preserve examples of animals both with mineralized skeletons and soft-bodied species lacking skeletons.

Among the fossils found in the Burgess and Maotianshan shales is the proto-arthropod *Anomalocaris*. This has been dubbed the 'great white shark' of the Cambrian oceans. At a time when most marine animals were small, *Anomalocaris* may have grown up to 38 centimetres (1 foot 3 inches) in length. However, for almost 100 years after its discovery the large size of *Anomalocaris* was not appreciated. The story begins with the description in 1892 by Joseph Whiteaves (1835–1909) of what was believed to be the segmented body of an average-sized fossil shrimp, *Anomalocaris* – meaning 'anomalous shrimp'. Twenty years later another peculiar fossil – *Peytoia* – resembling a pineapple ring and thought to be a jellyfish, was discovered in the Burgess Shale by Charles Doolittle Walcott (1850–1927) of the Smithsonian Institution. A third fossil – *Laggania* – seemed to be a sea cucumber, a worm or even a sponge, so odd were its features. In the 1980s a team of palaeontologists based at the University of Cambridge revealed the link between *Anomalocaris*, *Peytoia* and *Laggania*. Remarkably, they are all parts of the same or very similar animals, with *Anomalocaris* representing the paired appendages near the front end of these giant arthropods, *Peytoia* the mouthparts, and *Laggania* the main part of the body.

Anomalocaris turns out to be one of several predators belonging to the arthropod order Radiodonta that swam in the Cambrian and Ordovician seas. They were equipped with large stalked, compound eyes, each containing more than 15,000 lenses, endowing them with the sharp vision required for hunting prey visually. Their large, spine-laden frontal appendages and mouthparts are additional evidence of a predatory lifestyle. Although it is tempting to view these animals as biting predators of trilobites and other shelled animals, new studies of the mouthparts suggest they may have been relatively weak and acted by sucking soft food into their mouths. Perhaps the 'great white shark of the Cambrian oceans' was more like a modern flounder, feeding on prey inhabiting the mud and sand of the seabed.

Worms from sea to land
Hallucigenia

HALLUCIGENIA – THE NAME SAYS IT ALL! This small animal looks like the product of a surrealist's dream, a Cambrian equivalent of *The Elephant Celebes* painting by German artist Max Ernst (1891–1976). The name *Hallucigenia* was coined in 1977 by University of Cambridge palaeontologist Simon Conway Morris (1951–), although the fossil had first been found in the Burgess Shale of British Columbia, Canada in the early years of the twentieth century. Conway Morris noted that no animal like *Hallucigenia* was alive today. Based on the flattened fossils, he produced a reconstruction depicting a peculiar creature with a long body and seven pairs of stilt-like structures supporting it above the muddy seabed. At one end was a globular head and there were tentacles arranged along the top of the body in a single line. Conway Morris assumed *Hallucigenia* to be an immobile inhabitant of the seabed, using the tentacles with their divided tips to grasp particles of food and perhaps to pass them on to the mouth.

Within 15 years of Conway Morris's publication, Swedish and Chinese palaeontologists had discovered better-preserved specimens of *Hallucigenia* from Chengjiang in southwest China that literally overturned the original reconstruction of this animal. The stilts on the underside of the body were actually spines on its back. An additional row of 'tentacles' was found, which together with the first row could now be interpreted as paired legs on the under-side of the animal. The new caterpillar-like reconstruction of *Hallucigenia* made much more biological sense: the legs enabled locomotion, while the spines on the back had a protective function. But to which phylum of animals did *Hallucigenia* actually belong? The most compelling similarities are with Onychophora, the velvet worms. Not only is the bodyshape of *Hallucigenia* very like that of a modern velvet worm, but the divided tips of the 'tentacles' match the tiny claws found at the tips of the legs of living velvet worms such as *Peripatus*.

Although seldom much in evidence, there are about 180 species of modern velvet worms, living either in the tropics or countries south of the equator. All are terrestrial animals, hunting mostly insects at night and growing up to 20 centimetres (8 inches) long. By contrast, *Hallucigenia* lived in the sea, as did various related 'lobopods' found in rocks of Palaeozoic age, and grew to a mere 5.5 centimetres (just over two inches) long. At some unknown time in the history of onychophorans, these animals colonized the land and the marine forms subsequently became extinct. Few, if any, other animal phyla have made such a dramatic and complete switch in habitat during their evolutionary history.

Early echinoderms
cinctans

THE OLDER THE FOSSIL, the harder it is to interpret. This 'rule' certainly applies to many of the fossil animals that populated seas of the Cambrian period. Some of these fossils have such a strange appearance that it is very difficult to know to which modern group they are most closely related. Cinctans fall into this category of problematical fossils.

Resembling miniature tennis rackets just a few centimetres in length, cinctans are rare fossils found only in rocks of Middle Cambrian age. They have skeletons consisting of a series of plates. A ring of plates around the border of the fossil forms the cinctus from which the animals take their name. Within this ring are smaller polygonal plates, like the webbed strings of a tennis racket. A handle-like appendage extends from one end of the animal and is also constructed of plates. Not much else is evident apart from a large opening at the end of the fossil opposite to the appendage and a second, smaller opening. This opening is believed to be the mouth through which the living animal swallowed small food particles, although it is difficult to discount conclusively the alternative interpretation that it is the anus.

What might these strange fossils represent? The best clue comes from the fine structure of the plates: each individual plate comprises a highly porous crystal of the mineral calcite. Although calcite skeletons are found in many different groups of animals, including molluscs (p.28) and brachiopods (p. 48), only the echinoderms (starfish, feather stars, sea cucumbers, sea urchins and sea lilies) have plates with such a porous structure. But cinctans are asymmetrical animals whereas echinoderms today exhibit five-fold symmetry, as is so clearly apparent in starfishes with five arms. Indeed, pentaradial symmetry is regarded as a unique defining character of echinoderms.

So, cinctans appear to be echinoderms, yet they lacked the pentaradial symmetry found in echinoderms living in the sea today. In fact, cinctans represent an early stage in the evolution of echinoderms that preserved the ancestral asymmetrical symmetry of the most ancient echinoderms before pentaradial symmetry had evolved. Thus, the radial symmetry seen in modern echinoderms is an evolutionarily advanced feature, which contrasts with the primitive radial symmetry of, for example, sea anemones and jellyfish.

Palaeozoic plankton
graptolites

KNOWN COLLOQUIALLY AS 'the writing in the rocks', graptolites are peculiar colonial animals found abundantly in rocks of Early Palaeozoic age. Like pencil marks, graptolite fossils are often made of carbon, formed by the degradation of their original skeletons of collagen. Some graptolite colonies possess a single branch, others two or more branches, which are often curved or spiralled. The two-branched colonies of *Didymograptus* shown opposite resemble tuning forks but with the addition of serrated outer edges.

Until the 1970s it was unclear how graptolites were related to modern animals, even though they had been intensively studied by palaeontologists because of their value in dating Palaeozoic rocks. However, a link was eventually made between graptolites and some living animals called pterobranch hemichordates. Both pterobranchs and graptolites have colonies of numerous individual cup-like thecae – the serrations visible in the graptolite *Didymograptus* correspond to the locations of these individual thecae. In pterobranchs each theca is occupied by a zooid with a feathery, two-armed tentacle crown for feeding on tiny plankton. We must assume a similar anatomy in graptolites. A key piece of evidence uniting graptolites with pterobranchs is the way in which the skeleton is constructed from bandage-like strips of material applied to the outside of the thecae.

Although pterobranchs can help palaeontologists understand what a living graptolite was like, graptolites are far more diverse and show much greater morphological variety than living pterobranchs. Furthermore, while all pterobranchs live on the seabed, the majority of graptolites were planktonic animals floating or swimming in the open sea. It is not easy to find an exact ecological equivalent to graptolites among modern, large-sized planktonic animals. Perhaps the best candidates are salp chains, colonial sea squirts forming long strings of barrel-like zooids. However, both the zooids (up to 10 centimetres/4 inches) and the colonies (up to 20 metres/65 feet 7 inches) are considerably larger than graptolites. Salp chains can also wind around like a string of beads, whereas graptolite colonies were relatively rigid, capable of no more than slight flexing.

Fierce debates have raged about whether graptolites floated passively at the whim of the ocean currents, or were able to propel themselves actively through Palaeozoic seas. Turning to living salp chains for guidance, the pumping of water through the zooids to extract planktonic food results in a kind of jet propulsion novement. Possibly the feeding currents of graptolite colonies had a similar effect, but we may never know for sure. With the extinction of the planktonic graptolites in the Early Devonian, colonial macroplankton disappeared from the fossil record forever.

Cousins of starfish
edrioasteroids

YOU COULD BE FORGIVEN for thinking that this object is a Celtic stone carving. However, the circular motifs with their swirling patterns are, in fact, fossils of animals called edrioasteroids that in this case lived attached to the surface of another extinct marine animal known as a conulariid. Edrioasteroids are cousins of modern starfishes and sea urchins, whereas the conulariid is the sea-floor-dwelling stage of a kind of jellyfish (see p. 56). In both cases the original mineralized skeletons of the animals have been dissolved by water passing through the rock. This has left the fossil as a mould in the fine sandstone rock, reinforcing the erroneous impression that it is a carving.

The clue that edrioasteroids are echinoderms lies in their five-fold symmetry (see p. 22). Edrioasteroids have been described as looking like small starfishes sitting on top of a cushion composed of numerous small plates. The five arm-like structures, known as ambulacra, contain plates that could apparently be opened. This allowed the animal to feed on plankton, possibly using tube-feet similar to those found in living starfish. Beneath each ambulacrum is a groove that conveyed food to the mouth at the centre of the animal. In some species of edrioasteroids the ambulacra are straight, but in others they are curved – as in the species shown, which was collected from Ordovician rocks in Morocco. Curved ambulacra can have an anticlockwise or a clockwise curvature, and sometimes one of the five is curved in the opposite direction to the other four.

How did edrioasteroids live? The majority of edrioasteroid fossils are found attached to the surfaces of other fossils, such as the conulariid that hosted the Moroccan examples, or to brachiopod shells. In many instances evidence implies that the edrioasteroids colonized living animals, possibly to the detriment of the host. Other edrioasteroids occur on rocky submarine platforms, often in very dense populations with neighbouring animals touching one another. There is debate about whether edrioasteroids could move around or were permanently cemented in one place throughout their life. Regardless, they may be viewed as the Palaeozoic ecological equivalents of the acorn barnacles (p. 188) that also cover hard surfaces and which are particularly common on rocky shorelines today.

Ordovician predators
Orthoceras

NAUTILOIDS HAVE A LONG and illustrious fossil record, extending back nearly half a billion years. During this time they have given rise to all the other cephalopod molluscs, including the squid, cuttlefish, octopus and the celebrated ammonites, which they have famously outlived. Cephalopod molluscs all share some striking features, including large eyes, a head full of tentacles and a beaked mouth. The living pearly nautilus shows us that the tentacles and the beak are used to predate smaller animals. The tentacles grapple and hold the prey while the nautilus ejects toxic digestive juices, and then the beaked mouth crushes and devours it. It is highly likely that early nautiloids, such as *Orthoceras*, consumed their prey in much the same way, but unlike the rather meek shrimp-eating present-day nautilus, some of them evolved to reach enormous sizes, up to 6 metres (19 feet 8 inches) in length. Such massive tentacles and robust beaked mouth must have presented a formidable, kraken-like opponent to the sea scorpions, the 50-centimetre (20-inch) long trilobites and even other nautiloids that joined them in the Ordovician seas 450 million years ago. This was the nautiloids' heyday. Jawed fishes had yet to evolve and the nautiloids majestically filled the role of the top predator.

Modern and younger nautiloids have a coiled shell (p. 88), but fossils show us that most Palaeozoic nautiloid species had a straight or orthoconical shell like that of *Orthoceras*. When these orthoconical fossils are polished, the internal chambers that would have been gas filled buoyancy aids are handsomely revealed, and they are frequently sold in fossil shops, often with many shells together on a single slab. After death, the gas-filled chambers in modern nautiloids keep the shells afloat so that they often drift far and wide to be washed up on exotic beaches distant from their natural habitat. It is likely that a similar thing happened to fossil nautiloids. Rocks composed of almost nothing but orthocones may be evidence of these sorts of accumulations.

All modern cephalopod molluscs inhabit pure seawater, and it appears that none of the nautiloids, ammonites, octopuses or squid ever evolved to take advantage of life in freshwater. Other molluscs, such as the snails, have wonderfully exploited freshwater and even conquered the land, and why the cephalopods haven't done so remains one of the many unsolved evolutionary puzzles.

Early symbiosis
crinoids and platyceratids

THE MODERN WORLD is full of symbioses – different species living closely together in varying degrees of mutual dependence. Sometimes the relationship is harmonious and both partners benefit; this is called mutualism. In other symbioses one partner benefits at the expense of the other; this is called parasitism. Unfortunately, very few symbioses in the geological past fossilize both of the partners. One that does is the remarkable symbiosis between crinoids (sea lilies) and platyceratid snails, a relationship that endured for 200 million years, from the Ordovician to the Permian.

The limpet-like platyceratids are invariably found clinging to crinoids. They are usually located on the calyx, that part of the crinoid containing most of the organs, including the gut and the gonads. A circular scar is sometimes visible on a calyx after a platyceratid has dropped off, the scar representing a thickening of the crinoid skeleton that grew around the position aperture of the platyceratid shell.

The exact nature of the symbiosis is a matter of continuing debate among palaeontologists. At least some platyceratids are positioned close to the anus of the crinoid host, leading to the theory that they were coprophagous, feeding on the waste material of the crinoid. Stealing the food of the crinoid before ingestion is another more palatable possibility. Other examples appear to have drilled a hole through the crinoid's skeleton, allowing access to the organs within the calyx. This suggests parasitism, perhaps on the meaty gonads of the crinoids. One study compared the sizes of crinoids that had platyceratids attached to them with those barren of symbionts. Had the association been beneficial to the host crinoids, those with platyceratids should be larger on average than those without. However, just the opposite was found – crinoids with platyceratids tended to be smaller. Being 'married' to a platyceratid was evidently disadvantageous to these particular crinoids, pointing to a parasitic relationship.

What does this particular symbiosis tell us about the long-term evolution of symbioses? Despite the uncertain nature of the interactions between the two partners – parasitic, mutualistic or a mixture of the two – it shows that symbioses can persist through vast intervals of geological time with little apparent change. If the parasitism hypothesis is correct, crinoids must either have tolerated the platyceratids, or been unable to evolve a way of ridding themselves of the parasites. Along with many other groups of animals, those groups of crinoids hosting platyceratids became extinct at the end of the Permian period. And by throwing their lot in with these crinoids, platyceratids themselves apparently met their own demise.

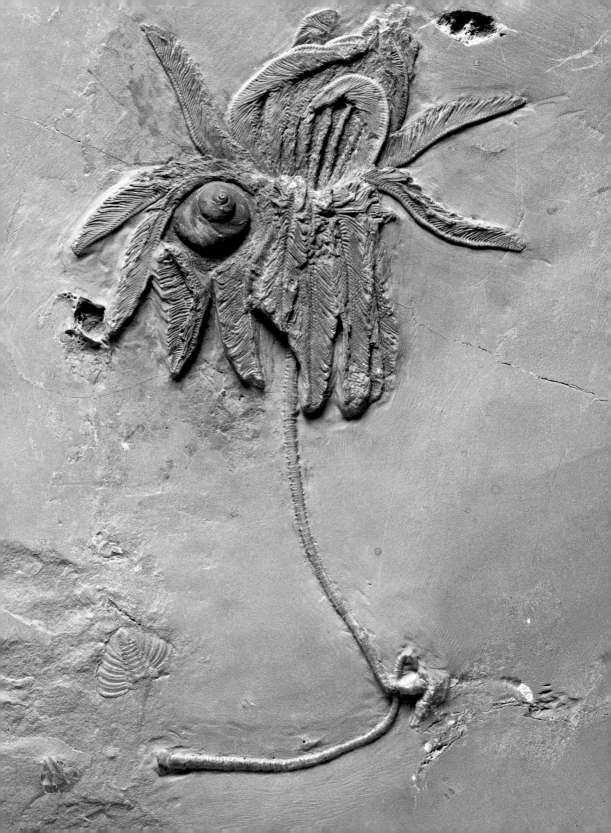

Mystery teeth
conodonts

IF YOU PLACE A BLOCK of limestone of Palaeozoic age into dilute acid until all of the soluble calcium carbonate has dissolved away, you may be fortunate to find an insoluble residue containing distinctive tooth-like microfossils. These are conodonts, first discovered and named in 1856 by the Latvian-born scientist Heinz Christian Pander (1794–1865). Typically less than one millimetre (1/32 inch) in length, conodonts possess either a single, curved conical cusp or a serrated edge consisting of numerous cusps arranged in a row. The reason that conodonts survive acid treatment is that they are made of insoluble calcium phosphate.

Palaeontologists studying conodonts in the early years of the twentieth century found them to be immensely valuable for dating rocks. Yet nobody really knew what kind of animal they represented, and a few dissenters even believed them to be plants. The discovery of occasional clusters of variously shaped conodonts arranged regularly in a pair of rows placed some constraints on interpretation. Nevertheless, the numerous reconstructions of the conodont animal, some more fanciful than others, could not be evaluated in the absence of a clear modern analogue and the biological identity of these microfossils remained enigmatic. The breakthrough came in the early 1980s with the discovery of a remarkable soft-bodied fossil in Carboniferous rocks near Edinburgh, Scotland. This fossil showed a group of conodonts near the end of an eel-shaped animal about 40 mm (1½ inches) long by 2 mm (1/16 inch) wide. A subsequent discovery in South Africa was made of an older (Ordovician) and much larger conodont animal. Both fossils show that the conodont animal had a head with a pair of eyes, and a segmented body.

The majority of palaeontologists now agree that most, if not all, conodonts represent the teeth of a chordate. Chordates are animals with a notochord, a stiff rod of cartilage that is replaced by the spinal column in the vertebrates. The existence of a head places conodonts among the craniates, slightly more advanced than lampreys and hagfishes, both of which lack any mineralized hard parts, but still primitive compared with the majority of chordates.

Conodont animals were abundant in the seas of the Palaeozoic, dwindling on into the Early Jurassic after surviving the great extinction at the end of the Permian period some 250 million years ago (p. 70). Little is known about their ecology, although they were probably good swimmers. New research has shed light on the function of the conodonts themselves.

As might be expected from their resemblance with teeth, they almost certainly functioned as teeth and were used for grasping, handling and slicing prey. The sharpness of conodonts is unparalleled among teeth, and this compensates for their small size and the limited force they could apply when feeding. Using the high magnifications afforded by electron microscopes, it is often possible to see minute scratches on the surface of conodonts. These microwear patterns resemble those visible on some mammal teeth and represent scour marks made when eating tough food items.

Ruling reefs
Halysites

THERE ARE MANY IMPRESSIVE structures on Earth that overwhelm astronauts viewing them from space. Seen as swathes of alluring turquoise waters fringing land masses and oceanic islands in the tropics, these reefs are formed by calcium carbonate secreted by living animals and plants in the shallow warm waters of the tropics. Reefs are built up over thousands of years by the growth of animals and plants. Most of the reefs we see today are constructed by a single group of animals called the scleractinian corals (p. 208). However, scleractinian corals have not always been in control of the valuable real estate provided by shallow tropical waters. The fossil record reveals how, for nearly two billion years, reefs have been built by a variety of different animals and plants.

The earliest reefs were simple structures, formed by the layering of billions of microscopic cyanobacteria that glued sand particles together to form large structures called stromatolites (p. 10). Then, when complex animal groups diversified, the stromatolites' dominion quickly ended. Two separate groups of corals, very unlike those that build reefs today, began jostling for supremacy. The rugose corals appeared in the Ordovician followed shortly by the tabulate corals, such as this beautiful 425-million-year-old chain-coral *Halysites* from the Silurian period. But their heyday wasn't to last, and these two coral groups both went extinct at the end of the Permian, 250 million years ago, in the largest global extinction event to have struck the Earth.

Recovery after the extinction during the Mesozoic era saw a struggle between a motley mix of skeletonized sponges, coralline algae and corals, all jostling for the role of ruling reef builder. The tussle continued until the entire world warmed significantly in the Cretaceous period, stifling most reef builders with intolerable heat. There was one group that could withstand the heat and take advantage of the vacant niche, however. Rudists (p. 136) were large bivalve molluscs that built magnificent barrier reefs hundreds of kilometres long.

The rudists were themselves driven to extinction along with the ammonites and dinosaurs at the Cretaceous–Tertiary (K–T) boundary 65 million years ago; the Earth cooled and the scleractinians' supremacy of the reefs returned. Despite times of global warming, global cooling and mass extinctions, the scleractinian corals have held the position of chief tropical reef builder ever since. Only now are the scleractinian corals in peril. Reefs are shrinking from an onslaught of human influences, ranging from pollution to ocean acidification, and some have died completely. After two billion years of hegemony, we are at risk of losing the turquoise tropical seas that so astound the distant astronauts as they marvel at the beauty of the Earth.

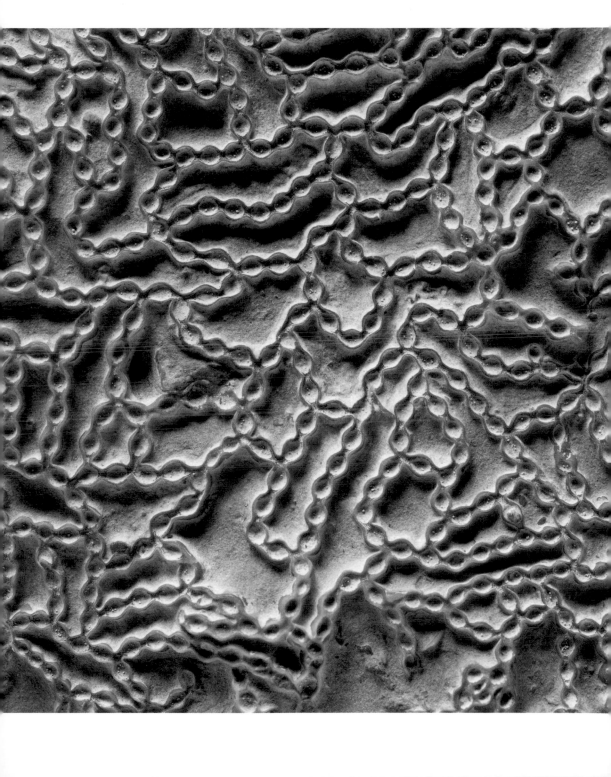

Forests of stone
sea-lilies

GIANT KELP FORESTS are among the most spectacular ecosystems in coastal seas today. Measured from their holdfasts anchored to rocks on the seabed to the photosynthetic fronds extending upwards to the surface of the water, the seaweeds at the heart of these diverse communities can be up to 50 metres (164 feet) in length. Lacking hard parts, kelp fronds have no fossil record, but a different kind of submarine forest that in all probability rivalled the splendour of modern kelp forests is exceedingly well represented in the fossil record: the stony submarine forests of sea-lilies. And like modern kelp forests, ancient sea-lily forests provided habitats for myriad other creatures.

Despite their name, sea-lilies are animals. Known more formally as stemmed crinoids, they are close relatives of starfish and sea urchins, and classified within the phylum Echinodermata. Their stems are anchored to the sea floor by holdfasts not unlike those of kelps. The crinoid stem functions as a tether for the 'business end' of the animal, consisting of a head-like crown containing the gut and reproductive organs, plus numerous arms – the 'petals' of the lily. Whereas kelp fronds intercept light for photosynthesis, crinoid arms intercept plankton for food. Remarkably, examples are known among the relatively few sea-lilies in modern oceans of stems continuing to live for significant periods after decapitation of the crown and arms. The stem remaining can apparently subsist by absorbing dissolved organic matter from the seawater.

Skeletons of individual crinoids are constructed of enormous numbers of calcareous elements held together by soft tissues. For instance, the stem, which on occasions could grow up to 40 metres in length (over 130 feet), is made from a stack of coin-shaped elements, each with a hole in the centre like a doughnut. These disaggregate and become scattered after the death of the animal. Dense forests of crinoids living in the Palaeozoic produced immense quantities of calcareous debris that in time accumulated to form crinoidal limestones. The greatest deposits of crinoidal limestones are of Carboniferous age, when forests of sea-lilies carpeted the shallow seas that spread across the continents.

There is an interesting parallel between the demise of sea-lily forests after their heyday in the Carboniferous and the decline of giant kelp forests today. Predators are implicated in both cases. Tethered sea-lilies became fixed bait for newly evolved predators in late Palaeozoic seas. Some sea-lilies survived by migrating to the refuge of the deep sea, where predators were less common. Others evolved to lose the stem and become mobile, using their arms to swim

away from would-be predators. Indeed, the crinoids surviving today are mostly stem-less 'feather stars', many living as nocturnal inhabitants of coral reefs. Giant kelp forests today are in retreat because of the dual effects of climate change and predators, especially grazing sea urchins that have increased in numbers as humans have caught the fish that previously fed on them. By contrast, not only are the changes affecting modern kelp forests happening too quickly for adaptation, but the kelp plants do not have the option of moving into the darkness of the deep sea or swimming away from their predators.

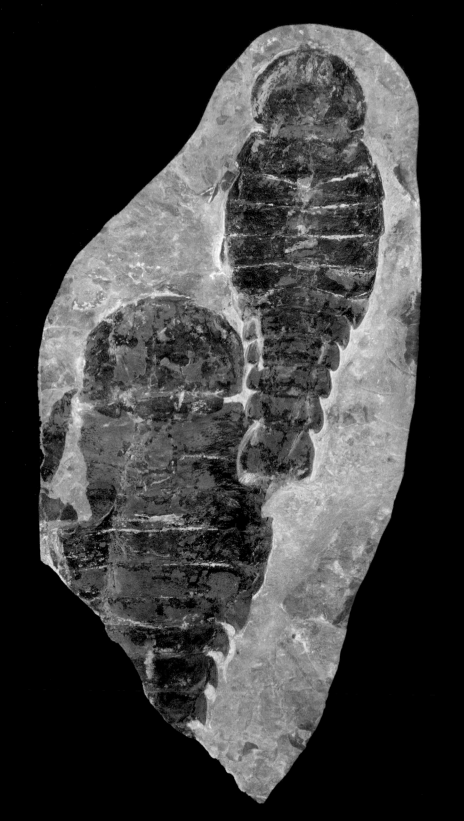

Ancient scorpions
Pterygotus

ARTHROPODS ARE ANIMALS with segmented bodies, rigid external skeletons and jointed limbs and they are by far the most diverse and abundant animal group in the world today. It is estimated that three-quarters of all living animals are arthropods, and there are an astonishing five million species, although some believe the true number to be double that.

Although the sting of a scorpion or wasp, or bite from a spider, is often painful, it is rarely life threatening. Fortunately, most arthropods are fairly small, and we are, comparatively, fairly large. But some 460 million years ago in the Ordovician period the eurypterids (sea scorpions), appeared and although some were a similar size to modern scorpions, others grew to become monsters over 2 metres (6½ feet) long, some of the largest known arthropods that ever lived.

Eurypterids looked very much like their namesake, with a flattened body, a pair of large grasping claws and a spine at the end of a segmented tail. Although they are related to present-day scorpions they actually represent an earlier evolutionary trajectory and pre-date true scorpions by around 30 million years. The earliest eurypterids are all represented by fossils preserved in ancient sea sediments that betray their salty origins. The majority of later examples are found in lake sediments, indicating that they shifted and adapted to a freshwater habitat.

Pterygotus was a predator and a scavenger, taking advantage of pretty much anything that came its way, including juveniles of the same species. Its spiny front limbs were perfect for grasping and killing writhing prey, and its spiny tail was an efficient armament. Unlike its modern relatives, the tail contained no venom. It didn't need it. It was a deft swimmer too, and it had to be, as larger species probably preyed upon the smaller ones. By pushing its two large paddles through the water, much like the aquatic beetle, the water boatman (common backswimmer), it could achieve rapid acceleration to catch prey or escape becoming the quarry of another. Despite being the top predator of its time, eurypterids are not too difficult to find as fossils if one looks in the right rocks, as most of the finds are actually moults rather than the animal itself. Like all other arthropods, eurypterids had to shed their tough exoskeleton to allow their soft body to grow, and so one individual could produce many fossils. In some rocks, the huge numbers of fossil moults found suggest that, just like their cousins the horseshoe crabs, eurypterids congregated in mass mating and moulting events. The sea scorpions went extinct during the Permian–Triassic mass extinction event 250 million years ago, since when there have been equally ferocious reptiles and mammals, but none quite so creepy or crawly.

Land grab
Cooksonia

NAMED AFTER INFLUENTIAL Australian palaeobotanist Isabel Clifton Cookson (1893–1973), the small *Cooksonia* is an archetypal fossil that showcases one of the most momentous events in the history of life. Over four hundred million years ago vascular plants made the evolutionary leap from the seas to the land – a position they have dominated ever since.

Before this decisive incursion, the sea had proved to be a highly successful cradle of life. Swathes of diversity had already evolved into many different ecological niches, including photosynthetic plants, filter-feeding fish and huge, predatory sea scorpions (p. 38). In stark contrast, the land was a relative biological desert, ready to surrender to whichever group could take advantage of its inhospitable terrain. Yet, the vascular plants weren't the first to attempt the conquest. There is good evidence that mosses, lichens and bacteria were present before *Cooksonia* flourished, but none were able to succeed in the way that *Cooksonia* did.

Cooksonia was a simple branching plant just a few centimetres tall. Although it had no leaves, it possessed several unique features that helped it overcome the most challenging aspect of living out of the water: the formidable threat of desiccation. Microscopic analysis of *Cooksonia* shows it clearly has a functional vascular system – a way of transporting water and nutrients from the soil up through the stem, thus ensuring that the critical parts of the plant were constantly fed with water and helping it survive hot dry spells. Remains of a waxy cuticle can also be seen. All modern plants use cuticle to prevent water escaping from their leaves. However, while a waterproof cuticle is extremely effective in keeping the water inside the plant, it also stops gases from the air moving into the plant's tissues. To continue to 'breathe' it was therefore essential that another decisive feature evolved concurrently. Stomata are small holes in the plant cuticle wall that all vascular plants possess. They permit the transfer of gases between the air and the cells. Without them plants would be unable to grow to hundreds of feet in height or survive desert-like conditions. And there they are in the fossil *Cooksonia*.

Each small stem of *Cooksonia* ended in a trumpet-shaped structure called a sporangium, which was used to spread the plant's spores as far and wide as possible, helping them colonize the surrounding territory. Small plants like *Cooksonia* dominated the land for 40 million years, setting the stage for more complex plants and eventually the arrival of terrestrial animals by stabilizing soils and oxygenating the atmosphere. The plants had conquered the land and staked their claim.

Life sucks
Cephalaspis

BEFORE JAWS EVOLVED, fish had to make do with sucking, squeezing and pulsating motions of their muscular mouths to devour their prey. *Cephalaspis* was a 25-centimetre (10-inch) long, distinctly shaped jawless fish that lived in freshwaters during the early part of the Devonian period, around 400 million years ago.

The mouth of *Cephalaspis* sits on the underside of its head, like that of a stingray, suggesting that like the stingrays it fed from the muddy sea floor. There is also evidence that the animal possessed some form of sensory organ on its underside that would have been used to detect food lying shallowly buried in the sea-floor sediment.

Without jaws its food could not have been particularly tough to ingest. Instead the diet of *Cephalaspis* probably consisted of soft-bodied animals such as shrimp and worms, and the fine particles of food to be found in detritus. The animal's most conspicuous feature is its armoured head and two fearsome-looking 'cornua' that projected backwards from the head. The cornua may have conferred protection from predators, although it is hard to see how they might have defended the animal. It is instead more likely that they served to churn up the sea-floor sediment to uncover buried prey, as some modern fish do, but no one knows for sure.

Its well-protected head was clearly a form of protection from predators, but what preyed upon *Cephalaspis*? Without conclusive evidence of predation it is always difficult to say, but possible predators were the massive sea scorpions – eurypterids (see p. 38), and the heavily plated placoderms (p. 53) – jawed fishes that had recently appeared on the scene and that clearly had the evolutionary upper hand. The threat appears to have been real, as the two eyes of *Cephalaspis* are set close together and centrally on the top of its head – perfectly located to alert the fish to anything looming above while it scoured the sea floor for morsels.

Early vision
trilobite

MOST OF US TAKE VISION very much for granted. But when did eyes first evolve? This is not a straightforward question because eyes appeared progressively, starting from microorganisms with simple receptors that could distinguish only between light and dark. However, the oldest fossils with complex eyes possessing lenses for focusing an image date from about 540 million years ago. Some evolutionary biologists interpret the advent of complex eyes as a driving force in the early proliferation of multicellular animals in the sea, the so-called Cambrian Explosion (p. 18). The basic argument is that active predation became possible only after eyes had evolved, forcing prey species to find new ways of not being detected and triggering a cascade of evolutionary responses between predators and prey.

Fortunately, the eyes of one particular group involved in the Cambrian Explosion had lenses made of calcite, a resistant mineral that fossilizes extremely well. Trilobites, an extinct group of arthropods, possessed compound eyes composed of numerous – up to 15,000 – small polygonal lenses. In three dimensions these lenses have the form of short cylinders, each capable of focusing a beam of light onto a photoreceptor surface at the back of the eye.

There are few finer examples of trilobite eyes than those of *Erbenochile*, a distinctive genus from the Devonian of Morocco and Algeria that lived about 400 million years ago. This trilobite is unusual in that the two eyes are like small towers. About 500 lenses cover the surface of each eye, and a projecting ledge caps the eye itself. The ledge may have functioned as a kind of sun shield, reducing glare for this animal, which inhabited shallow seas. *Erbenochile* is effectively a trilobite wearing a sun hat. Taken together, the two eyes gave *Erbenochile* 360° vision. Individuals were apparently capable of detecting very small movements, which would have had great value in detecting prey and predators alike. One possibility is that *Erbenochile* lived partly buried with only the eyes protruding above the sediment, in much the same way as some modern flatfish do today.

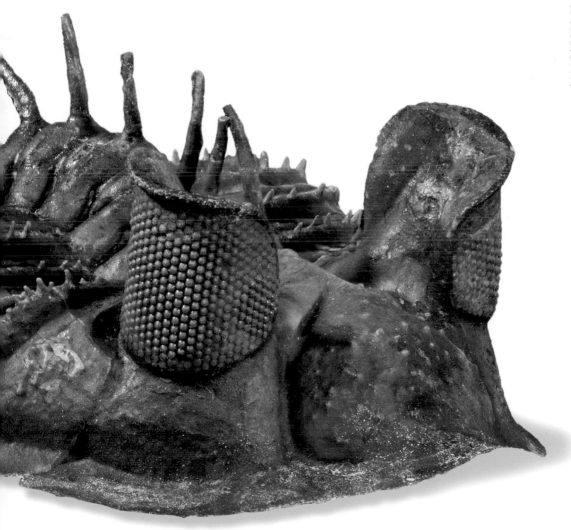

Avoiding predators
spiny trilobite *Comura*

THE THEORY OF EVOLUTION by natural selection put forward by Charles Darwin (1809–1882) is often paraphrased 'survival of the fittest'. Organisms have to survive in order to reproduce and leave offspring for future generations. A prerequisite for survival is not to be eaten by a predator. Interactions between predators and their prey have undoubtedly been an important force shaping the history of life, with predators constantly required to evolve new structures and behaviours in response to the evolving anti-predatory structures and behaviours of their prey.

Many predators in the sea are able to crush the shells of their prey. The existence of such 'durophagy' in the geological past can be seen from characteristic patterns of shell breakage in fossils, but more often palaeontologists must seek evidence from skeletal structures that can be interpreted as adaptations to counteract predators. Pointed spines are arranged over the

upper surfaces of many trilobites, as in this exquisite Moroccan specimen of *Comura* from the Devonian period. There is every reason to believe that the spines of trilobites functioned principally to protect them against predators. Trilobites were also able to discourage predators by rolling up into a ball, thereby enclosing the vulnerable underside of the animal. Enrolled spiny trilobites resemble rolled-up hedgehogs. The prominent spines of the enrolled trilobite must have presented a challenging impediment to potential assailants, including durophagous predators.

Tracing the fossil records of marine durophagous predators has shown how many groups were either absent or rare in the Cambrian and Ordovician periods. The subsequent blossoming of these predators brought about a change called the 'Middle Palaeozoic Marine Revolution'. The new durophagous predators include several kinds of fishes, crustaceans and eurypterids (p. 38). Potential prey animals responded by evolving thicker shells that were harder to crush, as well as defensive spines. Spines evolved not only in trilobites but also proliferated in brachiopods (p. 48) and crinoids (p. 36) at this time. However, spines could never be an ultimate deterrent in this evolutionary arms race in the face of the appearance of increasingly larger and more powerful predators capable of crushing prey, spines and all.

Compete or fail
brachiopods

INVERTEBRATE ANIMALS WITH shells of chalky calcium carbonate are the mainstays of the fossil record. Countless species of shelled animals have populated the sea since Cambrian times, many living a largely or completely immobile existence on the sea floor, filtering plankton from the water around them. Two completely unrelated groups of such animals have a body protected by a pair of hinged shells able to open slightly to admit plankton-laden seawater. These are the bivalve molluscs, including mussels and scallops, and the far less familiar brachiopods.

Brachiopods, known by the vernacular term 'lamp shells' because of their vague resemblance to Roman lamps, were the dominant group of twin-shelled animals for much of geological time. They come in a variety of shapes and can be distinguished from the majority of bivalve molluscs, as they have shells of unequal size and shape: one of the shells is larger than the other and overarches it above the hingeline, where there is often a round opening. A stalk, attaching the animal to a rock or sediment on the seabed, emerges from this opening in the living animal. Brachiopods were enormously abundant in Palaeozoic seas from about 500 until 250 million years ago. However, fewer brachiopods are encountered in younger rocks, and today they are abundant only in some parts of the southern hemisphere, including Antarctica. At the same time as brachiopods declined in the fossil record, bivalve molluscs became more abundant. In view of their generally similar appearance and co-occurrence in the same habitats, this led to the suggestion that the bivalve molluscs and brachiopods competed, with the dominant bivalve molluscs winning out and slowly replacing the brachiopods.

In the world of human commerce, competition has driven many companies into decline and eventual extinction. But has competition had the same impact in animals and plants over long periods of geological time? The fossil record provides few potential candidates for such competitive replacement, brachiopods and bivalve molluscs being the most widely heralded example since Swiss palaeontologist Louis Agassiz (1807–1873) first suggested it back in 1857. Modern brachiopods do indeed seem to be more sluggish feeders and can fare badly in the presence of bivalve molluscs. However, in a 1980 paper, 'Ships that pass in the night', American palaeontologist and evolutionary biologist Stephen Jay Gould (1941–2002) attributed the decline of brachiopods relative to bivalve molluscs to the failure of brachiopods to recover from the great extinction at the end of the Palaeozoic (p. 71) and not to competition with bivalve molluscs. The jury is still out, but, as with so many historical events, we may never have a definitive explanation.

Jaws
Cheiracanthus

SOME FEATURES OF AN ANIMAL or plant are so efficient at their task that they would appear to have been purposefully designed. The jaw is a perfect example. The ability to catch, handle and devour food with strength using articulated structures at the front of the head is one of the most important and successful traits of both insects and vertebrates, and it is difficult to imagine animal life without them. But far from being designed, the jaws of many insects began rather incredibly as front legs that over time became more effective at handling food than locomotion. Vertebrate jaws arose via a very different but equally fascinating and unlikely route, each step providing some advantage over its predecessor.

The gills of fish, except the jawless slime-producing hagfish, are held open by looped structures known as brachial arches, letting water pass freely through the gills and allowing the fish to efficiently 'breathe' water. Primitive jawless fish had muscular mouths unsupported by either bone or cartilage and, as is revealed by *Cephalaspis* (see p. 43), they probably fed using sucking and squeezing motions rather than biting. Fossils show that these early fish also possessed more brachial gill arches than their modern relatives. Over many generations the set of arches that sit closest to the mouth moved forward and became hinged, creating a prototype toothless jaw. The ability to firmly grasp prey that were otherwise prone to escape

was revolutionary and allowed these early vertebrates to greatly broaden their diet and take advantage of new sources of food that had appeared with the proliferation of complex and mobile life.

The acanthodians, more charmingly known as the spiny sharks, were one of the earliest groups of vertebrates to make use of this new-fangled device. Despite their name, spiny sharks were not closely related to true sharks, and indeed preceded them by 50 million years. However, they did possess some similarities, such as a streamlined body for fast swimming, a skeleton made of cartilage, a blunt strong head and an upturned tail. Unlike sharks, their fins were supported by bony spines – hence their name.

The fine example illustrated here is an acanthodian called *Cheiracanthus murchisoni* preserved in rock that is found in present-day Scotland. Now long extinct, *Cheiracanthus* flourished 385 million years ago in what was then warm tropical freshwater lakes and rivers teeming with life. Although considerably much smaller than the famous white shark in the film *Jaws*, and only having the lower jaw fused with teeth, *Cheiracanthus* must have similarly used its jaw as an efficient and ruthless killing machine.

Shark's nemesis
placoderms

IN EVOLUTIONARY TERMS sharks have done pretty well for themselves. Since their humble origins in the Silurian period, 430 million years ago, sharks have seen a number of setbacks, including rapid climatic changes, global mass extinctions and the arrival of stiff competition from reptiles and mammals that returned to the seas. Yet the sharks have rebounded time and time again. Today they are considered the top predators in many of the world's oceans.

At the dawn of this tumultuous history, the sharks' nemeses were placoderms. These were a group of extinct fishes known only from their fossils, like the one shown here, discovered in Scottish Devonian rocks. The placoderms ruled the oceans for fifty million years and are well known to palaeontologists, thanks to the thick bony scales that covered their head and thorax, which have been preserved so well.

Both sharks and placoderms originated in the Silurian period, but while the sharks remained a small and rare group of fishes, the placoderms went on an evolutionary rollercoaster that shared many similarities with the ultimate diversification of the sharks. Like the sharks of today, the placoderms took advantage of a wide range of habitats, including both saltwater and freshwater. They reached enormous sizes and without doubt held the role of top predator in the seas. They even took the more placid role of eating plankton, a role that is majestically exploited today by the basking and whale sharks. But the shark's time was yet to come; the Devonian was the heyday of the placoderms, and there is evidence that some made meals of the few paltry sharks around.

Even so, their supremacy wasn't permanent. At the end of the Devonian period a massive extinction rocked the placoderms and they disappeared entirely. But one group's loss was another's gain, and the sharks took advantage of the respite, diversifying rapidly into the vacant ecological roles left by the placoderms, and establishing themselves, at least temporarily, as the kings of the oceans.

Invading the land
Eusthenopteron

EUSTHENOPTERON IS A DEVONIAN lobe-finned fish – a fish with paired, lobe-like pelvic and pectoral fins, each joined to the body by a single bone. These bones are important in being the equivalents of the bony limbs found in land-dwelling tetrapods, including amphibians, reptiles, birds and mammals. Indeed, *Eusthenopteron* is described in many old books as a direct ancestor of the first amphibians. Confidently recognizing direct ancestors among fossils is tantamount to impossible, however, and *Eusthenopteron* should rather be regarded as a close cousin to some, as yet, unknown fish that was ancestral to the tetrapods.

The fish–tetrapod transition is one of the key events in the history of life on Earth, leading to the colonization of the land surface by vertebrates and ultimately to the evolution of humankind. Two major problems had to be solved for this change to occur: how to breathe oxygen from air rather than water; and how to convert locomotion by swimming to walking. Lungfish living today provide a vital clue to the evolution of air breathing. In these lobe-finned fishes, the swimbladder, used by other fishes to control buoyancy, is modified to allow air breathing. Revealingly, in most species of lungfish the gills are completely redundant and the fishes must breach the water surface in order to obtain the oxygen they need to live. As already mentioned, the bones in the pelvic and pectoral fins of lobe-finned fishes formed the basis for the evolution of the limbs of tetrapods. New fossil finds and interpretations show how bony limbs were first used for sculling around in shallow waters, only later becoming used for walking on the land surface.

Just how did *Eusthenopteron* live, and was there anything about its ecology linking it to that of the tetrapods? All indications are that *Eusthenopteron*, which could reach a length of 1.8 metres (5 feet 11 inches), was a completely aquatic animal that never emerged fully from the water. The locality of Miguasha in Quebec, Canada has yielded thousands of specimens of *Eusthenopteron*. Here, about 380 million years ago, *Eusthenopteron* apparently lived in an estuary.

The discovery of pieces of other fishes inside fossils of *Eusthenopteron* shows that it was a predator. Indeed, there is a general similarity in the overall shape of *Eusthenopteron* to modern pike, notably in the long body with fins concentrated towards the back end. Pike remain stationary in the water before ambushing their prey using a very rapid turn of speed – *Eusthenopteron* probably fed in much the same way. The earliest tetrapods inherited the predatory lifestyle seen in *Eusthenopteron* and other Devonian lobe-finned fishes. However, the suctorial feeding employed by fishes, including *Eusthenopteron*, does not work out of water. Tetrapods had to evolve other techniques, such as overtaking their prey and directly biting down onto it.

Square jellyfish
conulariids

JELLYFISH BELONG TO the same phylum – Cnidaria – as sea anemones, corals and *Hydra*. And like these other cnidarians, they possess stinging cells, as swimmers unfortunate enough to encounter jellyfish may have discovered. On the face of it, the soft body of a jellyfish should not make a good subject for fossilization. Indeed, examples of fossil jellyfish are notable for their rarity. However, fossils called conulariids, which are very common in some Palaeozoic and Triassic rocks, are believed to be a kind of jellyfish, despite their very different appearance.

Conulariids resemble ice-cream cones but have four-sided, rather than round, cross-sectional shapes. The largest measure up to 50 centimetres (20 inches) in length from the narrow apical end to the broad open end, although most are less than 10 centimetres (4 inches) long. Wavy ridges cross the surfaces of the cones, with the transverse bands usually disrupted or deflected along the mid-line of each of the four faces. When preserved, the shell material of conulariids is composed of calcium phosphate, like the bones of vertebrates but unlike most invertebrate animals.

Why are these extinct animals thought to be jellyfish? After all, jellyfish are not conical, and nor do they have skeletons of calcium phosphate. To find the answer we must look more closely at jellyfish. Like most cnidarians, jellyfish have two alternating stages in their life cycle, a medusoid and a polypoid stage. The medusoid stage is the familiar free-swimming jellyfish. Medusoids are budded from the far less conspicuous polypoid stage. The latter has tentacles and resembles a sea anemone resting on the seabed. Crucially, the polypoid stages of jellyfish show a distinctive four-fold symmetry, just like conulariids. This is good evidence that conulariids represent the polypoid stage of jellyfish. The corresponding medusoid stage of these extinct jellyfish, if one existed at all, has yet to be recognized in the fossil record.

In the polypoid stages of some modern jellyfish, the soft parts are protected within a tube. Although the tube is never reinforced by calcium phosphate in species living today, conulariids did apparently evolve biomineralized skeletons. It is not unusual for fossils to reveal features present in the ancient relatives of modern groups that have been lost completely in their living descendants. The question can be posed as to whether this capability is still latent somewhere within the genes of modern jellyfish, or was lost forever with the extinction of conulariids over 200 million years ago during the Triassic period.

Forest fuel
Lepidodendron

COAL IS SYMBOLIC OF THE industrialization of the Western world and has been called 'black diamond' for the wealth it brought the owners of the land in which it was found. But what exactly is coal? Since the colonization of the land by plants over 400 million years ago, there have been many places in the world where the humidity is high and the land low and subsiding – perfect for a thick blanketing of swamp. Growth of trees in wet tropical forests is so prolific that rotting vegetation piles up quickly. If the underlying rock is sinking or the sea level rising, layers upon layers of dead trees and rotting vegetation are laid down and preserved under yet more organic material and sometimes incursions from the sea. After burial, the vegetation becomes compressed, buried further and heated over millions of years, causing carbonization – or in other terms, the production of coal. Around 330 million years ago conditions were so conducive to the formation of coal by this process that the period was named the 'Carboniferous'. At this time Britain and much of North America were located in the tropics and had vast low-lying lands ideal for the growth of luxuriant wet forests.

The champion trees of these Carboniferous tropical forests were a group of early vascular plants called scale trees. *Lepidodendron* (Greek for 'scale tree'), shown opposite, was one of the most widespread and abundant of these majestic plants. It was also one of the largest, reaching heights of 40 metres (131 feet 3 inches) or more and having a trunk 2 metres (6 feet 6 inches) wide. Looking closely at the trunk, it's possible to see stomata and cuticle, revealing that the trunk was probably photosynthetic and would therefore have been green. Initial growth was upwards, rapid and unbranching. Only when this massive single shoot reached the canopy did the tree form a beautiful crown of branches tipped with cone-like structures, from which its spores could spread far and wide.

Lepidodendron was a major contributor to the coalfields of Europe and North America. It is these 330-million-year-old carbonized forests that fuelled the industrial revolution in the nineteenth and twentieth centuries, when coal was burnt to power railways, steamships and mills. Even today, most of the world's electricity comes from the burning of coal. However, the benefits of coal come at a price, and the detrimental effects of burning it were known long before the industrial revolution. Burning coal emits mercury, selenium, arsenic and sulphur into the atmosphere, the last of these returning as acid rain. Nearly 10,000 million tonnes of carbon dioxide is released by coal burning every year, making it one of the greatest drivers of human-induced climate change.

Continental drift

Glossopteris

WHEN THE RESCUE PARTY RECOVERED the sledge used by Robert Falcon Scott's team on the ill-fated Antarctic Terra Nova expedition of 1912, it contained the leaves of a fossil plant, *Glossopteris*. The expedition's scientist Edward Wilson (1872–1912) had collected the fossils and, although he may not have known their scientific significance, the fact that he did not discard them on the desperate journey back from the South Pole is testimony to his high regard for their value.

Geologists in the late nineteenth century were struck by the geographical distribution of *Glossopteris* and the other land plant fossils with which it characteristically occurs. This '*Glossopteris* flora' is found in rocks of Permian age in South America, southern Africa, Madagascar, India and Australia – as well as Antarctica. However, elsewhere in the world the *Glossopteris* flora is totally absent. At first this distribution was explained by the former existence of 'land bridges' linking land masses that are today separated by sea. Austrian geologist Eduard Suess (1831–1914) proposed the name Gondwanaland for the hypothetical southern land mass containing the *Glossopteris* flora. Later, the idea of former land bridges was abandoned and the *Glossopteris* flora became a key piece of evidence supporting the theory of continental drift. Gondwanaland was, in fact, an ancient continent split apart by continental drift, leaving fragments in what are now South America, southern Africa, Madagascar, India, Australia and Antarctica. In the 1960s it was established that continental drift is brought about by slow movements of the Earth's crustal plates (plate tectonics) over millions of years – movements that caused the eventual break-up of Gondwanaland long after the *Glossopteris* flora had lived.

Rather little is known about *Glossopteris* despite being a major constituent of economically important coals mined in India. Its relationship to other plants is uncertain, although it was clearly a type of gymnosperm (a seed-producing plant). It was a woody plant, with a trunk up to 80 centimetres (31½ inches) in diameter, suggesting that complete trees could attain 30 metres (98 feet 5 inches) in height. These trees formed forests in swampy mires, much like the older forests of the Carboniferous (p. 59). An important difference, however, is that the Carboniferous coal swamps were steamy forests located close to the ancient equator, whereas the Permian *Glossopteris* coal swamps could exist at much higher latitudes. Indeed, the *Glossopteris* fossils found by Scott's expedition in the Antarctic probably grew at a latitude of over 80°S, similar to the present-day latitude where they were found. No polar ice was present when the plants lived, but they still had to be able to survive long periods of winter darkness within the Antarctic Circle.

Giant amphibians
Eryops

AMPHIBIANS TODAY TEND TO BE relatively small and often inconspicuous animals, although the largest living species – the critically endangered Chinese giant salamander – can occasionally grow up 1.8 metres (5 feet 11 inches) in length. In Palaeozoic times, giant amphibians were more commonplace. The largest, the crocodile-shaped *Prionosuchus*, reached an estimated length of 9 metres (29 feet 6 inches). The skull shown here comes from another huge Palaeozoic amphibian, *Eryops*, which has been found in the Early Permian rocks of Texas and New Mexico, USA.

Eryops could grow up to 2.5 metres (8 feet 2 inches) long. Fossils of *Eryops* found in Texas come from sediments that formed in swampy environments with lush vegetation. The animals probably divided their time between the water and the land: their short, powerful limbs suggest they were more than capable of sustained walking on the land surface. The front limbs had four fingers and the back limbs five, as is true for most amphibians, fossil and living. The tail was moderately long, slightly over one-third of the total length of the animal. Relative to the body, *Eryops* had a large skull that was broad and flat, and a very wide mouth. The numerous sharp teeth show that it was a predator, although possibly not a particularly mobile one. Fish, terrestrial animals, or a mixture of both, may have formed its diet.

Like *Prionosuchus*, *Eryops* belongs to the dominant group of Palaeozoic amphibians called temnospondyls, which were once thought to be reptiles. Some temnospondyls possess scales and an armour-plated body, features never found among modern amphibians and more often associated with reptiles. Temnospondyls first appeared about 330 million years ago in the Early Carboniferous, and evolved into a great variety of different forms adapted for aquatic, semiaquatic and terrestrial lifestyles. Even the most terrestrially adapted temnospondyls had to return to water to lay their eggs. Their larval stages would have inhabited lakes and rivers before metamorphosing into adults. A larval amphibian called *Onchiodon* from Germany may have been very similar to the larval stage of *Eryops*, although none are known from Texas. Temnospondyls survived the Permian mass extinction and lived on into the Mesozoic, before finally becoming extinct in the Cretaceous. However, mounting evidence points to temnospondyls being the ancestors of the lissamphibians, the biological group that includes all living species of amphibians. If so, the genetic legacy of temnospondyls may still be with us in the amphibians of today.

Whorls of teeth
Helicoprion

THE INFINITE MONKEY THEOREM posits that a monkey, given an infinite amount of time and provided with a typewriter, would eventually produce the complete works of Shakespeare. Evolution sometimes seems to operate like this. Random events (the mutation of genes) during seemingly infinite time (3.5 billion years of life on Earth) have undoubtedly produced many weird and wonderful creatures. But in reality evolution doesn't work like that.

Take *Helicoprion* – a unique fossil found in marine rocks from around 290 until 270 million years ago. It has a spiral, or whorl, of what appear to be teeth, sometimes 150 or more. Although the 'teeth' are very shark-like, there's absolutely nothing like it in the natural world today, and palaeontologists have been perplexed by it for over 120 years. Reconstructions have ranged from a shark with a coiled toothed snout, a serrated coiled tail, a toothed dorsal fin and even an extendable toothed tongue. Recent high-resolution computed tomography (CT) scans of *Helicoprion* have moved palaeontologists towards finally resolving the conundrum by forming 3D reconstructions of the creature's cartilaginous head. The tooth whorl sat within the lower jaw and grew, tooth by tooth, with the largest, more recent ones exposed at the front. As larger teeth were added, the smaller ones were pushed back inside the jaw, where they coiled up and were kept for life. When the beast closed its mouth it would have created a circular slicing action. For all its grisliness however, it appears that this formidable weapon wasn't used on anything more robust than a paltry shrimp, as none of the teeth found show the wear expected if *Helicoprion* were feeding on animals with tough skeletons. It turns out the creature was not even a true shark, but a distant relative.

Was this evolutionarily novel jaw the result of random, infinite-monkey-type evolution? In 2003, a group of students placed a computer with a keyboard in the macaque enclosure of Paignton Zoo for a month. The monkeys used the machine as a lavatory and bashed it with rocks, and yet the experiment was not entirely devoid of scientific worth. The monkeys did manage to type out five pages of text, but most of it consisted of the letter 'S'. Something about the letter 'S' obviously appealed. The monkeys' manuscript was not in fact random at all, and neither is evolution. Like all of life, the fantastical teeth of *Helicoprion* resulted from the incremental acquisition of features, each of which must have been successful. Although the genetic mutations that led to the arrival of the teeth were random, the survival of each increment leading up to them was not. Many strange and inexplicable fossils only appear to be so outlandish that only randomness could explain their evolution simply because they are extinct and challenge our frame of reference.

Sex or solar panels?
Dimetrodon

ANY YOUNGSTER INTERESTED in palaeontology will tell you that contrary to its appearance *Dimetrodon* was not in fact a dinosaur but lived 50 million years before the first dinosaurs appeared. Dimetrodon wasn't even an ancestor of dinosaurs but was closely related to the therapsids, the branch of land vertebrates that included the precursors of all mammals (p. 77).

This beast must nonetheless have been a fearsome top predator. This 275-million-year-old *Dimetrodon grandis* was 5 metres (16 feet 5 inches) long and had a jaw lined with formidable rows of serrated teeth and piercing canines near the front. It had a powerful tail, sturdy, muscular legs and a powerful bite that would have efficiently killed, cut and consumed its prey. This level of proficiency had become necessary because by the time *Dimetrodon* had become top predator, its principal prey, such as large amphibians and fish, had themselves become agile escape artists. For all its terrifying appearance, the most striking features of *Dimetrodon* are the greatly elongated spines on the vertebrae. What was the purpose of this sail-like structure along its back? Early palaeontologists suggested that the sail may have provided camouflage in long reeds, given strength to the backbone, or even caught the wind like a true sail and pushed the animal through water!

Another much examined explanation is that the sail was used by *Dimetrodon* to regulate its body temperature. Undeniably, a cold-blooded animal of such large size would have taken a long time to heat up to optimal temperature simply by basking like a lizard. Its smaller prey would certainly have had the advantage. To speed up the process *Dimetrodon* could have turned its sail to face the sun, heating the circulating blood and hence the whole body. Likewise, the sail could have played a role in cooling the animal down and preventing overheating. It is an appealing theory but of the twelve or so named *Dimetrodon* species, most are much smaller than the arresting *Dimetrodon grandis* and would not have required anything like the level of thermoregulation needed by this large monster, yet they too have well-developed sails. What's more, estimates suggest that it may have taken up to 4 hours for *Dimetrodon grandis* to heat up using the sail method, making the sail's usefulness as a thermoregulator somewhat suspicious. Another group of palaeontologists argue that the sail was instead a brightly coloured ornament used to attract mates or perhaps to intimidate rivals. The larger and more flamboyant the sail the greater chance of successfully mating, thus leading to the 'sexual selection' of larger and ever more ostentatious sails. The truth is likely to be a fusion of both theories, as such multitasking is a common theme in evolution. The vibrant but slightly ludicrous beak of the toucan is used to attract mates, and yet it also functions well to pick fruit and snatch chicks from other's nests and to regulate body temperature.

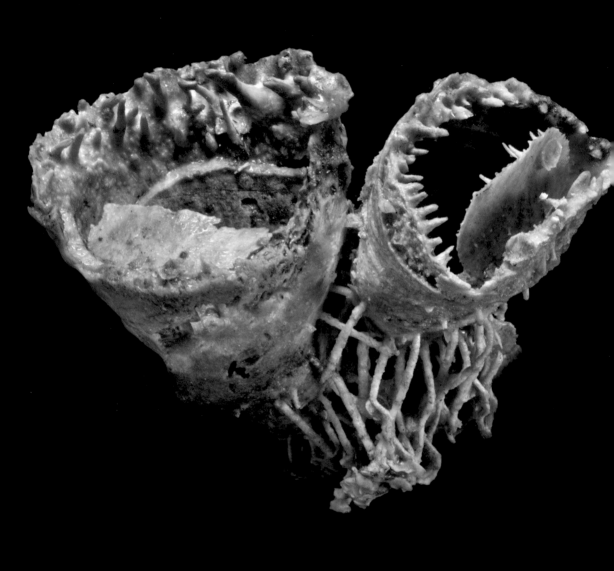

Preserved by silica
Cyclacantharia

THE TROUBLE WITH FOSSILS is that many are so firmly embedded in very hard rocks that it can take countless hours carefully chipping them away from the rock. Nature sometimes does the job for us – surface weathering may gradually free any fossils that are more resistant than the enclosing rock. Fortunately, another natural phenomenon – silicification – opens the way for acid extraction of almost perfectly preserved fossils from limestones. Some of the most exquisite Palaeozoic fossil shells are preserved by silicification, such as the bizarre richtofenid brachiopod *Cyclacantharia*. Shells originally made of calcium carbonate can be replaced by various silica minerals, including quartz and a fine intergrowth of quartz and moganite, known as chalcedony. The process of silicification occurs over a long period, during which the shell is progressively replaced by silica. Often, this silica originates from the dissolution of the siliceous skeletons of other fossils. When silicified fossils are etched in acid to free them from the surrounding limestone, the faithfully replicated fine details are revealed.

One of the most famous silicified fossil faunas comes from the Glass Mountains of Texas. American palaeobiologists G Arthur Cooper (1902–2000) and Richard E Grant (1927–1995) immersed blocks of Permian limestone from the Glass Mountains in hydrochloric acid. They recovered wonderfully preserved fossils of marine animals belonging to hundreds of species that populated a shallow sea. Most of the brachiopods were conventional in appearance – they looked essentially like clams with the two valves of the shell almost, although not exactly, mirror images of one another (see p. 48). By contrast, richtofenids superficially resembled corals with a large cone-shaped valve accommodating a much smaller and very different second valve.

Richtofenids are an example of an aberrant group – one that differs radically from the norm. The evolution of these animals took a different course as they became adapted to an unusual lifestyle for a brachiopod. The large conical valve was cemented firmly to the seabed, its stability helped by the growth of prop-like spines, unlike most brachiopods in which a fleshy pedicle tethers the shell and allows some degree of movement. It has been suggested by some palaeontologists that the small valve flapped rhythmically to suck water into the conical large valve, where the feeding organs were located. But this idea has been justifiably criticized; it is more likely that the small valve simply opened when the animal was feeding and closed to protect the vulnerable soft parts of the animal when danger threatened. Whatever its function, without serendipitous silicification we would know far less about *Cyclacantharia* and a host of other fossils from hard, uncompromising limestones.

Mass extinctions
blastoids

THIS PERMIAN FOSSIL, *Deltoblastus*, from the Indonesian island of Timor is one of the last of its kind. It belongs to a class of echinoderms called blastoids that had prospered in the oceans of the world for some 200 million years. The fate that befell blastoids at the end of the Permian period was shared by numerous animals in the sea, as well as plants and animals on the land surface. This was the time of the 'mother of all mass extinctions', when life on Earth came within a whisker of complete oblivion. It has been estimated that 95% of marine species disappeared during the Permian mass extinction event, compared with only 75% during the younger end-Cretaceous mass extinction when the dinosaurs met their demise (pp. 138–139). When such a large proportion of species becomes extinct, it is inevitable that some entire biological groups will disappear too. Joining the blastoids in the culling of life at the end of the Permian were trilobites, rostroconch molluscs, fenestrate bryozoans and rugose corals. All had been important components of Palaeozoic ecosystems in the sea.

What was the cause of the Permian mass extinction? Unlike the end-Cretaceous mass extinction, there is no strong evidence for an asteroid impact, and terrestrial causes for the catastrophe are more likely. Current thinking favours a train of events that raised the temperature on the Earth's surface rapidly and made the oceans stagnant. This sequence could have been ignited by the eruption of massive volcanoes, now represented by two million square kilometres (780,000 square miles) of lava in eastern Russia, called the Siberian Traps. The huge volumes of the greenhouse gas carbon dioxide released into the atmosphere would have elevated global temperatures. Some scientists believe that the hotter climate would have destabilized methane normally locked up as 'gas hydrates' in sediments beneath the world's oceans, causing release of this even more powerful greenhouse gas and a further rise in temperature. The conveyor-belt like circulation pattern of water in the oceans may have ceased, as the warmer surface waters of polar seas would no longer have been dense enough to sink. The ensuing stagnation of bottom waters would have been to the obvious detriment of marine life on the seabed.

A remarkable feature of the Permian mass extinction is its long-term effect on the Earth's biosphere. Although there are some notable exceptions, rocks of the succeeding Triassic period are typically poor in fossils, and those fossils that do occur tend to be small and simple (p. 74). It took well over 50 million years for the diversity of marine animals to return to the pre-extinction level of the Permian. Part of the reason for this slow recovery may be the continuing stresses faced by animals and plants living in the post-extinction world of the Triassic. However, it is also possible that organisms and ecosystems were inherently unable to rebound quickly from this catastrophic mass extinction, a potent lesson from the past for those considering the human-propelled mass extinction we are currently experiencing.

MESOZOIC ERA

POSITIONED BETWEEN TWO mass extinctions, the Mesozoic era lasted for almost 200 million years and is subdivided into three periods: Triassic, Jurassic and Cretaceous. For geologists, this is a fascinating era, during which the great continent of Pangaea split into Gondwana in the south and Laurasia in the north, separated by an equatorial ocean called Tethys. Fragmentation of Gondwana stranded numerous land-dwelling animals and plants, some representing survivors from more ancient times. The Mesozoic was also the time when the Atlantic Ocean formed, gradually dividing the Americas from Europe and Africa. There were few or no polar ice caps during the Mesozoic, which reflects warmer global temperatures, and also higher sea levels than we have today.

The Mesozoic is often referred to as the 'Age of Reptiles'. This was the era of dinosaurs, as well as their flying cousins the pterosaurs, and in the sea the dolphin-like ichthyosaurs, plesiosaurs and mosasaurs. Crocodiles and turtles added to the panoply of reptiles, while bony, ray-finned fishes first became diverse. The first birds appeared and mammals and flowering plants both made their debuts.

Invertebrate life continued to evolve in the sea. Decimated by the mass extinction at the end of the Permian, a Lilliputian world of small-sized invertebrates is believed by some scientists to have characterized Early Triassic seas. Recovery was slow, but by the Jurassic two groups of swimming cephalopod molluscs had appeared – ammonites and belemnites – to be followed in the Cretaceous by their cousins, the octopuses. Reefs were constructed not only by corals, but also by some peculiar bivalve molluscs that flourished in the warm waters of the Tethys Ocean. Various predatory animals appeared, making life ever more difficult for their prey and driving the evolution of better methods of defence. Much of the evolution of marine animals in the Late Cretaceous was played out in the Chalk Sea that stretched across the continental shelves of the northern hemisphere. Here a rain of microscopic plates of algae blanketed the sea floor on which animals such as sponges lived.

Mesozoic ecosystems collapsed at the end of the Cretaceous period during the famous 'K–Pg mass extinction' ('K' referring to Kreide, German for Chalk, and 'Pg' Palaeogene, the stratigraphical period that followed the Cretaceous). On land the last of the dinosaurs disappeared and in the sea the final species of ammonites perished. We are still far from understanding the complete story of the K–Pg mass extinction and questions remain regarding the rapidity of the extinction and its cause. The most favoured theory is that the impact of an asteroid off the coast of present-day Mexico was responsible, another that the massive volcanoes that created the Deccan Traps of India drove the environmental changes that brought about species extinctions. As with other mass extinctions, many species were wiped out, opening the way for other species to prosper and in the case of mammals, to trigger a major evolutionary radiation that was to follow in the Cenozoic Era.

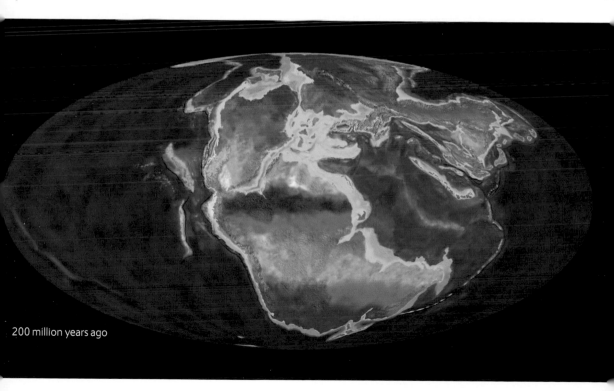

200 million years ago

The Lilliput effect
tiny Triassic snails

THE POST-APOCALYPTIC Earth of the future is invariably envisaged as one of devastated landscapes, with little life and scarce resources for the few animals and plants that do remain. Similar scenes may have pertained immediately after any of the 'big five' mass extinctions of the last half a billion years. Species of animal and plant fortunate enough to survive these catastrophes would have faced severe challenges if they were to persist. The environmental effects of whatever caused the mass extinction may have lingered on, and food would have been in short supply once the carrion had been consumed. Life would have been hard. The prevalence of small-sized fossils in sedimentary rocks formed immediately after mass extinctions – the so-called 'Lilliput effect' – is interpreted by some scientists as a reflection of the harsh conditions.

Why did small animals and plants flourish immediately after mass extinctions? The Lilliput effect can be explained in several different ways. First, all of the largest species may have suffered extinction. This is because populations of these species generally contain fewer individuals, making them more vulnerable to extinction than small-sized species that typically have larger, and often more widely dispersed, populations. With species of large body size gone, the average size of fossils in post-extinction communities would have been reduced. Secondly, a world with a shortage of food may have suited small organisms better than large. Likewise, the lower concentration of dissolved oxygen in the sea, which is thought to have characterized these troubled times, would have favoured small animals. Simple geometry shows that small size brings with it a relatively larger surface area compared to volume, which would have helped to absorb the limited oxygen available to animals living in the sea.

However, not all scientists agree about the Lilliput effect. Some believe that, if present at all, Lilliput effects endured for only a short time in geological terms. Early claims were made that a Lilliput effect following the catastrophic mass extinction at the end of the Permian period lasted for a full five million years, well into the Triassic period, but new findings point to an effect no longer than 2 million years. Many Early Triassic fossils are, indeed, rather smaller than average, like the snails depicted here in a thin section of rock, but more research is clearly required before we can be sure about the existence of Lilliputian worlds following mass extinction events.

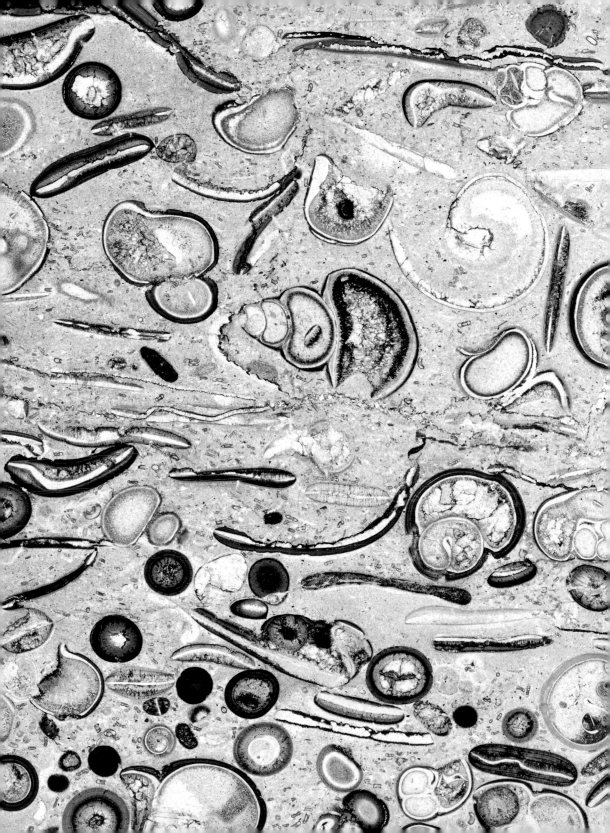

Halfway to mammals
cynodonts

HUMANS ARE JUST ONE OF 5,000 species of mammal living today. Mammals are enormously variable and have evolved adaptations to many different lifestyles, not only on land but also in the air (bats) and in the sea (whales, dolphins and seals). The origin of mammals can be traced through a group of mammal-like reptiles called synapsids to a particular kind of animal that lived during the Triassic period. These advanced synapsids, the cynodonts, are often considered to be textbook examples of evolutionary missing links: many features of the fossilized skeletons of cynodonts are distinctly reptilian, but the teeth, in particular, look like those of mammals. Mammal teeth are specialized for different functions: incisors for cutting, canines for stabbing, and molars for crushing and grinding food. Reptiles, by contrast, have an array of more uniform teeth.

The herbivorous cynodont *Massetognathus* had incisors, canines and flat-topped molars covered by low ridges, apparently an adaptation for grinding tough plant stems and roots. Fossils of *Massetognathus* have been found in Brazil and Argentina, occurring in Middle Triassic rocks about 235 million years old. This fox-sized animal had claws on its feet and a strong tail.

Did *Massetognathus* and other cynodonts possess other features regarded as typically mammalian? Did they have a hairy coat rather than scaly skin, give birth to live young rather than laying eggs, use milk to suckle their young, and have warm blood? The fossil record is usually mute on such questions if soft parts are not preserved. However, there is some evidence that cynodonts laid eggs, were warm blooded, as indicated by the detailed structure of the bones, and had a body covered by hair. The latter is suggested by the presence of structures interpreted as whisker pits in some bones on the snout. Cynodonts had a large braincase, more akin to those of mammals than reptiles. They also had a bony palate in the roof of their mouth, which would have separated the air passages to allow breathing at the same time as eating – a feat not possible in reptiles.

However, the structure of the cynodont brain, as determined from the shape of the braincase, implies poorer vision and sense of smell than the mammals that followed them. Cynodonts were indeed a halfway house between reptiles and mammals.

Start of something big
Megazostrodon

MAMMALS MADE AN INAUSPICIOUS debut about 225 million years ago, late in the Triassic period. For almost the next 160 million years, through the Jurassic and Cretaceous periods, they lived in the shadow of the dinosaurs. Most of the earliest mammals were small creatures. *Megazostrodon*, from the Early Jurassic of southern Africa, is a good example of one of these animals. First discovered in Lesotho in 1966, *Megazostrodon* weighed between 20 and 30 grams (¾ to 1 ounce) when alive. Its teeth show that it ate insects; it was possibly nocturnal and can be pictured rooting among leaf litter or foraging in trees for its food while the dominant dinosaurs slept. Compared with most modern mammals, the limb bones of *Megazostrodon* imply a more sprawling gait, with the legs emerging horizontally from the body. This reptile-like stance is less mechanically efficient than having limbs held directly beneath the body, as in modern mammals, and it seems likely that *Megazostrodon* could not scurry around with quite the same speed or dexterity as small mammals living today.

Why did it take mammals so long to become diverse and varied? The most popular explanation is that dinosaurs inhabited the major ecological niches that mammals later came to occupy. It required the extinction of dinosaurs to 'wipe the slate clean' and allow mammals to diversify into these niches. Only then did the apparently meek mammals inherit the Earth. However, the more we learn about Mesozoic mammals, the more niches they are found to have inhabited. There are now fossorial (digging), gliding and semiaquatic types, as well as the previously recognized ground- and tree-dwelling types. Many of these are known from skeletons with additional soft part anatomy, discovered in recent years in Early Cretaceous rocks in China. The largest weighed up to 14 kilograms (30 pounds 14 ounces) and fed on baby dinosaurs, the remains of which have been found in the stomach area.

New molecular methods of estimating the times of divergence between modern groups of mammals support this idea, finding that most of the evolutionary radiation in these mammal groups happened after dinosaur extinction, even though some of the major groups seem to have originated in the Cretaceous. The evolutionary radiation of mammals may have depended on a chance event well beyond the control of the animals themselves – catastrophic disruption of the Earth's biosphere caused by a colliding asteroid or volcanic eruptions of huge magnitude (see p. 139).

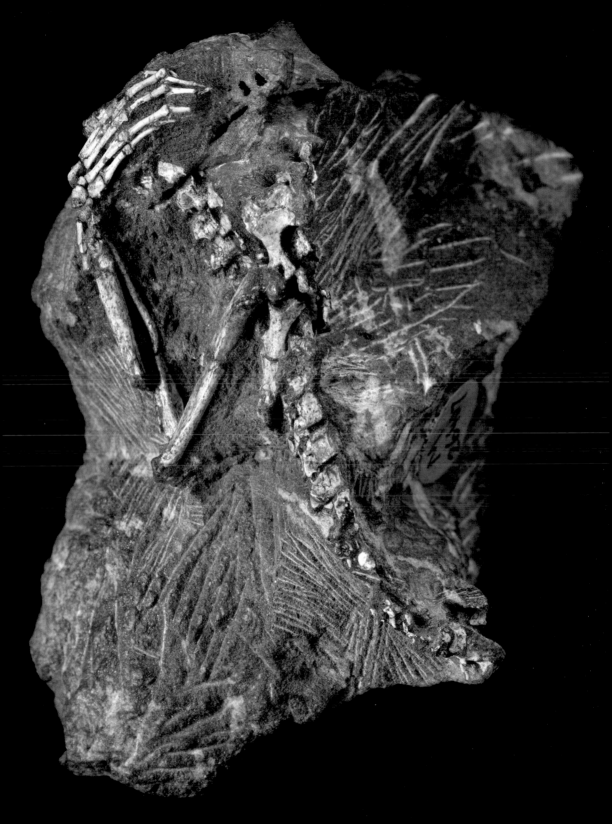

Safety in isolation
tuatara

SOME SPECIES CAN THANK geological changes for their survival. The tuatara, now living only in New Zealand, is a case in point. This creature survived because as New Zealand became isolated it was protected from the competition that brought about the extinction of relatives in other areas of the world.

Continental drift has had an important role to play in the diversification of life on our planet, not only by separating land masses and enabling them to evolve their own distinctive animals and plants, but also by isolating them from invasive species. The native animals and plants living today in New Zealand reflect its 80-million-year isolation from other land masses. Many of New Zealand's endemic species are refugees from the geological past, the last survivors of groups that once flourished globally. Perhaps the best example is the tuatara. Looking for all the world like a lizard, this small animal belongs to a different, more primitive group of reptiles, the sphenodontids.

Only one species of tuatara lives today. This greenish brown reptile has a spiny crest along the back and measure up to 80 centimetres (31 inches) long. Individuals can live for more than 100 years, are nocturnal and hibernate during the cold winter months. Once present on the mainland, tuatara are today confined

to about 32 small offshore islands, where they are protected from the predators introduced by humans to the mainland. In all, the total number of tuatara may be less than 100,000. But their sphenodontid relatives were far more common during the Mesozoic era, and not just in New Zealand.

Some of the best fossil sphenodontids have been found in the Mendip Hills of southwestern England and in South Wales. They occur in the sediments that infill fissures and caves in limestones. These fossils belong to animals that lived between about 220 and 200 million years ago, during the Late Triassic and earliest Jurassic periods. Among these is *Clevosaurus*, a small animal with a skull about 3 centimetres (just over one inch) in length, smaller than the New Zealand tuatara. Like all other sphenodontids, *Clevosaurus* differed from lizards in having two large openings on each side of the skull, whereas lizards have only one opening, and a small, overhanging beak on the upper jaw. The shape of the teeth in *Clevosaurus* suggests a diet of plants and insects. Nine species of *Clevosaurus* have so far been discovered – in Britain, Brazil, Canada and China – showing that this reptile had a much wider geographical distribution than the isolated tuatara does today.

Bony fish diversify
Dapedium

AMONG THE PAINTINGS OF Beatrix Potter (1866–1943), author of the *Peter Rabbit* books, is a watercolour depicting some fossils she had found in Gloucestershire, England. Potter was an enthusiastic naturalist and artist, as well as a writer of children's books. One of the fossils in her painting is a cluster of rhombic-shaped scales, representing a fragment of the fish *Dapedium*. Better-preserved, complete examples of this fish can be found in the Jurassic rocks of Lyme Regis, Dorset, coincidentally a known holiday retreat of Beatrix Potter. They include some magnificent specimens that were purchased by the British Museum in the 1800s during the early heyday of collecting from this prolific site.

Fossils of *Dapedium* reach the size of a large dinner plate. The fish has a shape reminiscent of the sea bream or porgy, although it is not a close relative of these living fishes. *Dapedium* has a body covered by thick, enamel-like scales. Blunt, pebble-like teeth fill the mouth. As in the sea bream, which has very similar teeth, *Dapedium* teeth were well adapted for crushing prey with hard shells, and it is likely that this fish fed mainly on molluscs.

Another fossil fish closely related to *Dapedium* is *Lepidotes*. It too has shiny, pebble-like teeth, which are found frequently as isolated fossils in Jurassic and Cretaceous rocks. In folklore, the teeth of *Lepidotes* are known as toadstones, reflecting an old belief that they represent the mythical stones once thought to be present in the heads of toads. Toadstone rings were common items of medieval jewellery. They were believed to sweat in the presence of poison, thus providing the wearer with an early warning.

Dapedium is an example of a bony, ray-finned fish. The great majority of fishes living today are of the ray-finned variety, the main exceptions being the sharks and their kin, which are cartilaginous and not bony, and the lobe-finned fishes, such as the coelacanth and lungfishes. Today there are some 27,000 species of ray-finned fish, making them the most diverse group of animals with backbones on our planet. They include such familiar, but disparate fish as carp, cod, plaice, trout and eels. Although ray-finned fishes have a fossil record extending back 425 million years to the Silurian, they did not become especially diverse until the Mesozoic. Newly evolved species adapted to fill different ecological niches in the sea and in freshwater rivers and lakes. However, the main evolutionary radiation of ray-finned fish occurred in the Cenozoic, when many of the families still living today first appeared.

Devil's toenails
Gryphaea

KNOWN IN FOLKLORE AS 'devil's toenails', *Gryphaea* shells can be found in great profusion in quarries and foreshore exposures of Early Jurassic rocks from Scotland to the south of France. The resemblance of *Gryphaea* to a gnarled talon accounts for its nickname, and the powdered shell of this contorted fossil was once used as folk medicine for ailments of the joints in horses and humans, the logic being that like could cure like. *Gryphaea* is, in fact, a bivalve mollusc, closely related to modern oysters. The shape of *Gryphaea* is particularly distinctive: one of the two valves is large, strongly convex and beak-like; the other is small, flat, lid-like and is almost hidden from view within the concavity on the inside of the large valve. This inequality contrasts with most other bivalve molluscs in which the two valves are more or less mirror images of each other.

Gryphaea provides us with one of the best examples of evolution in the fossil record. The pioneering research on the evolution of *Gryphaea* was undertaken by British palaeontologist Arthur Trueman (1894–1956) in the 1920s, but it was not until the later part of the twentieth century that the story became clearer thanks to the work of scientists on both sides of the Atlantic, including the renowned evolutionary biologist Stephen Jay Gould (1941–2002). This research showed how the oldest species, *Gryphaea arcuata*, the classic devil's toenail, evolved into *Gryphaea mccullochi*, which in turn evolved into *Gryphaea gigantea*. The entire evolutionary sequence took place over about 15 million years of geological time. It entailed an increase in the overall size of the mollusc, a decrease in shell curvature and the thinning, broadening and flattening of the shell. Despite debate over the tempo of these changes – did they occur at an even rate (gradualism) or were they stepped (punctuated)? – this evolutionary sequence has stood the test of re-examination by palaeontologists employing increasingly more sophisticated methods of analysis on ever larger collections of fossils. Furthermore, a similar sequence of evolutionary changes was repeated by *Bilobissa*, another *Gryphaea*-like oyster that appeared later in the Jurassic.

Why did both *Gryphaea* and *Bilobissa* evolve larger, broader and thinner shells through time? To answer this question we have to consider their mode of life. Whereas many oysters cement themselves to rocks or other shells, *Gryphaea* and *Bilobissa* adopted a prostrate lifestyle. The large convex valve of the animal rested partly buried in the mud on the seabed, with the gape between the two valves held above the sediment so that feeding on plankton from the seawater could occur above the mud. Maintaining a stable position was clearly vital to the survival of individuals of *Gryphaea* and *Bilobissa*, and this led to the evolution of lower, flatter and lighter shells that were less likely to sink in the sediment or be overturned than the high, narrow and heavy shells of the early species.

Mass mortality
Promicroceras

SOMETIMES FOSSILS ARE PRESENT in such prodigious numbers that it becomes difficult to see the rock for the fossils. A beautiful example is the 'Marston Magna Marble' from Somerset in England, a limestone packed with small ammonites. This is not a true marble, which is a metamorphic rock formed by the action of intense heat and pressure on limestone, but a hard sedimentary rock capable of being polished. The story goes that all known specimens of Marston Magna Marble, which is not today exposed on the surface, came from a well at a depth of 22 metres (72 feet 2 inches) beneath the surface. Some of the slabs of this ammonite graveyard were used, ironically, as headstones in graveyards.

Nearly all of the ammonites belong to a single species called *Promicroceras marstonense*. They indicate that the Marston Magna Marble dates from early in the Jurassic period, almost 200 million years ago. But why is the rock so packed with ammonites? There are two basic ways in which such fossil-rich rocks can be formed. The first is by the slow accumulation of organic remains over a long period of time without significant amounts of sediment to dilute them. The second is through the deaths of a large number of individuals at the same time in a mass mortality event. The pristine preservation of the ammonites in the Marston Magna Marble allows the first explanation to be discounted: ammonites that accumulated slowly are normally damaged, broken by currents and scavenging animals, corroded by seawater and colonized by encrusting organisms. The pale mineral infilling of the inner chambers of these ammonites shows that the shells remained unbroken after burial, preventing mud from entering. The only reasonable explanation of the Marston Magna ammonite graveyard is mass mortality.

Mass mortality events involving marine animals can happen for a variety of reasons, including starvation, disease and poisoning by water deficient in oxygen. However, another reason – spawning – is equally possible in the case of the Marston Magna Marble ammonites. We know very little about the reproductive behaviour of the extinct ammonites and must rely on information from their close relatives living today. The best analogues in this respect are squid. It is known that some species of squid congregate at restricted breeding grounds, mate, shed their fertilized eggs, and then die. The result is seasonal mass mortality at the squid spawning grounds. Perhaps Marston Magna was a spawning ground for ammonites 200 million years ago? As with so many questions in palaeontology concerning events from the distant geological past, we may never know the answer for sure.

Survivors
nautiloids

A JURASSIC NAUTILOID, *Cenoceras*, sliced in two and polished, shows the internal structure of the shell of this cephalopod mollusc. The shell describes a perfect logarithmic spiral, with later whorls becoming progressively broader as the growing animal added new shell material at the edge of the open aperture. Curved septa subdivide the shell into chambers, and are intersected by a tube crossing the chambers in the middle of the whorls. Most of this tube, which is known as a siphuncle, was filled by dark-coloured mud after the animal died, but as with the chambers, mud failed to infiltrate the entire structure. Those parts of the siphuncle and chambers not filled by mud contain the mineral calcite. This was precipitated from solutions passing through the sediment, in a similar way to the formation of a stalactite from water charged with minerals in solution. A feature of this particular specimen is the concentric banding of calcite, the first bands growing on the surfaces of the shell and septa, with later bands progressively closing up the chambers from the outside in.

Interpreting the biology of fossil nautiloids such as *Cenoceras* is made easier by the existence of several species of nautiloid living today in the warm waters of the Indo-Pacific. Not only can we reconstruct the main body of the animal occupying the outermost chamber, but we can also understand the purpose of the inner chambers and siphuncle. *Cenoceras* would have possessed a head with large eyes, indicating good visual capability, numerous short tentacles lacking the suckers found in squid and octopuses, and a tube (hyponome) through which water could be rapidly expelled to allow swimming by jet propulsion. The older chambers would have been filled in life by a mixture of gas and liquid. The siphuncle allowed these chambers to be accessed so that the balance of gas and liquid could be adjusted to allow the animal to rise or fall in the ocean.

Nautiloids coexisted with ammonites in Jurassic and Cretaceous seas. However, they were less abundant, contained far fewer species, and showed less variety in shell shape. Yet, except possibly for some local, short-lived populations, ammonites disappeared during the end-Cretaceous mass extinction, whereas nautiloids survived. How can this be explained? Were ammonites just unlucky, or did their biology differ in some crucial way from nautiloids, making them more vulnerable? Recent research on some Late Cretaceous ammonites found that they had jaws adapted for feeding on plankton. By contrast, to judge from modern species, nautiloids living at the same time would have consumed much larger food items. A major collapse of plankton in the sea is widely believed to have occurred at the end of the Cretaceous. So, maybe ammonites starved to death and became extinct, while the carrion-feeding nautiloids prospered and survived.

'Fish lizards'
ichthyosaurs

THE LEGENDARY EARLY BRITISH fossil hunter Mary Anning (1799–1847) made an extraordinary discovery in 1811. She found the skeleton of an animal 5 metres (16 feet 5 inches) long in the crumbling Jurassic cliffs between Lyme Regis and Charmouth in Dorset, England. This turned out to be the first complete skeleton of a distinct type of marine reptile that was subsequently given the name 'ichthyosaur' (fish lizard) by Charles Koenig (1774–1851) of the British Museum.

Ichthyosaurs were large predators that swam in the sea during the Triassic, Jurassic and Cretaceous periods. They had a streamlined body , shaped very much like that of modern dolphins, with limbs modified into fins for manoeuvring but a tail in the form of a vertical fluke, contrasting with the horizontal fluke of a dolphin's tail. Nevertheless, ichthyosaurs and dolphins provide an excellent example of convergent evolution. This is when two animals or plants have a very different ancestry but evolve to look similar. Another example of convergent evolution is afforded by birds and bats, two distinct types of flying animals sharing a basically similar appearance despite having descended from very dissimilar reptilian and mammalian ancestors, respectively.

Rather more than might be expected is known about the biology of ichthyosaurs. This is because several examples are exquisitely fossilized as a result of rapid burial of the carcasses in fetid muds that retarded decomposition of the soft parts. Indeed, some examples of ichthyosaurs from Holzmaden near Stuttgart in Germany preserve the outline of the fleshy tissues around the bones, allowing the dolphin-like shape of the body to be seen very clearly. Arm hooks of squid-like belemnites found in the body cavities of ichthyosaurs, where the stomach would have been located, provide evidence that the ichthyosaurs were active predators. Also remarkable are examples of adult ichthyosaurs containing embryos. In a few instances the females apparently died in the throes of giving birth. So, instead of laying eggs like most other reptiles, ichthyosaurs gave birth to live young while at sea. This is another attribute shared with dolphins; both animals are so well adapted to swimming in the oceans that they have lost the option of returning to the land to give birth or lay eggs – unlike turtles, for example.

Stealth and longevity
crocodiles

TICK-TOCK, THE CROCODILE IN Peter Pan that has swallowed a clock never quite gets his prey – Captain Hook. The clock's audible ticking sound prevents him from doing what crocodiles do best – sneaking up undetected before attacking their unsuspecting victims. Crocodiles are stealth predators and have probably hunted in much the same way since they first evolved in the Late Triassic, about 225 million years ago.

One of the few animals that dare venture close to modern crocodiles is the Egyptian plover or crocodile bird, which sometimes feeds on parasites living among the scales of Nile crocodiles. There is an irony here as, despite appearances, crocodiles and birds are closely related in evolutionary terms – they diverged from a common ancestor sometime in the Triassic period, evolving along very different trajectories and ending up looking utterly dissimilar. Crocodiles and birds are the two surviving representatives of the archosaurs, a group that also includes the extinct non-avian dinosaurs and pterosaurs.

Crocodiles, or more strictly crocodylomorphs – a large group including the 23 living species of crocodile, alligator and gharial, plus numerous related extinct forms – are represented in the fossil record by bones and teeth as well as bony scutes formed beneath the scales. The largest, *Sarcosuchus* from the Cretaceous, reached 12 metres (39 feet 4 inches) in length. This is three times longer than the biggest modern crocodile, the saltwater crocodile of Southeast Asia and northern Australia. Crocodiles have extremely powerful jaws able to exert a very high biting force, believed by some scientists to be important in explaining the great evolutionary longevity of these animals. Uniquely among modern crocodiles, the saltwater crocodile

may venture far from the coast into the open ocean. Similar marine crocodiles were, however, fairly common in the geological past. They included this *Plagiophthalmosuchus* from the Jurassic and Early Cretaceous. This crocodile was typically about 2–4 metres (6 feet 6 inches to 13 feet 1½ inches) long and had a slender body and sharp teeth, with a long snout resembling that of a modern gharial. *Steneosaurus* probably fed mainly on fishes, although examples of mollusc shells with puncture marks interpreted as made by this crocodile suggest that shellfish were also part of its diet.

Paddling into legend
plesiosaurs

THE PRESTIGIOUS SCIENTIFIC journal *Nature* published an article in its 11 December 1975 issue that received massive publicity. This was hardly surprising as the article gave a formal scientific name, *Nessiteras rhombopteryx*, to the legendary Loch Ness Monster. The publication was accompanied by a fuzzy photograph of a diamond-shaped paddle, adding to the library of indistinct images showing a long-necked, hump-backed animal that have led many to suspect that 'Nessie' is a plesiosaur. If true, it would extend the range of these large reptiles forward in time by 70 million years, as the known fossil record of plesiosaurs ends in the Cretaceous. But is the existence of a living plesiosaur in Loch Ness plausible? To answer this question we must summarize what is known about plesiosaur biology.

This is the paddle of a Jurassic plesiosaur. Despite vaguely resembling the long-necked sauropod dinosaurs, plesiosaurs belong to a completely different branch of the reptilian family tree and were fully aquatic marine animals, unlike the land-dwelling dinosaurs. The largest plesiosaurs reached 14 metres (45 feet 11 inches) in length, although the paddle depicted here belongs to an animal about a quarter of this length. Evidence from preserved stomach contents shows that plesiosaurs ate fishes and cephalopod molluscs, probably taking advantage of their long flexible neck to dart the head into position for

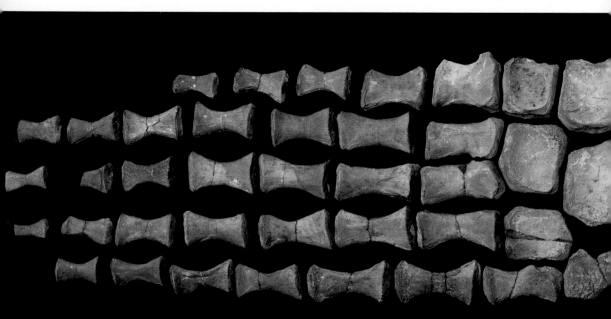

grabbing prey. The paddles acted as flippers, allowing the animals to swim slowly and manoeuvre themselves.

Some of the evidence for the existence of Nessie comes from sightings on land. Is it possible that these enormous creatures periodically emerged from the sea onto sandy beaches where they laid their eggs and perhaps, like turtles, used their flippers to dig holes in which to bury them? A recent find in Kansas of a pregnant adult plesiosaur with the bones of a single foetus in its body cavity shows this not to be the case. Instead of laying eggs, it seems that plesiosaurs gave birth to live young, as has long been known for ichthyosaurs, another group of large swimming reptiles that cohabited the seas with plesiosaurs during the Jurassic and Cretaceous (p. 90). Plesiosaurs, at least the long-necked species, were unable to venture onto land; without the buoyancy of the water, they would not have been able to support the weight of their own neck.

So, plesiosaurs could not have walked into the loch from the land. Loch Ness is a young lake. Until about 10,000 years ago this part of Scotland was buried deep beneath the ice. Therefore, Nessie's kin could only have colonized Loch Ness from the sea, but the geological record of marine sediments contains no trace of plesiosaur fossils for almost 70 million years. Then there is the problem of not just one but a viable population of plesiosaurs needing to be present in the loch. The hypothesis is not credible. Plesiosaurs are without doubt extinct, and our knowledge of these magnificent reptiles must therefore rely on the continuing study and discovery of fossils such as the Kansas specimen, and not on the stories of folklore, however appealing they might seem.

Colour and fossils
Neosolenopora

THE STENCH OF ROTTING SEAWEED cast ashore by seasonal storms is a routine annoyance on beaches around the world. It is, however, a reflection of the vast underwater meadows and forests of marine plants carpeting the seabed in shallow waters around the coast. While there can be no doubting that similar marine plants existed in the geological past, there is precious little evidence of them in the fossil record, which is hardly surprising as they lack readily fossilizable hard parts. But some types of marine plants secrete skeletons of calcium carbonate, and these can be preserved as fossils. Among the most striking of these is *Neosolenopora*. This plant, which belongs to an extinct group of algae, formed mound-shaped growths on the seabed. Examples from the Jurassic of Gloucestershire, England are pink to red in colour, giving rise to the vernacular name 'beetroot stone' for the limestone in which they occur.

Fossils seldom preserve their original colours, and hence the *Neosolenopora* fossils from the 'beetroot stone' have attracted scientific attention. The red pigmentation of these fossils is consistent with their identity as red algae, but the chemical responsible for the red colour has only recently been isolated. It turns out to be an organic compound called a borolithochrome, which contains, as the name suggests, the element boron. Surprisingly, modern red algae do not acquire their colour from borolithochromes, but instead from another type of organic compound, phycobiliproteins. The borolithochromes found in *Neosolenopora* are examples of molecular fossils. New techniques are enabling scientists to find ever more molecular fossils important in adding to our understanding of the history of life. For instance, molecular fossils in sedimentary rocks can indicate the former presence of plants and animals for which no conventional fossil remains are preserved. They also serve as biomarkers useful in petroleum exploration.

Another conspicuous feature of *Neosolenopora* is the banding visible in sectioned specimens. These alternating pale and dark pink bands clearly correspond to growth phases, just like tree rings. They are probably seasonal features, a couplet comprising one pale and one dark band constituting one year of growth by the plant. If this is correct, the largest specimens of *Neosolenopora* lived for about 20 years. The precise chemical composition of the skeleton even suggests that the pale bands were produced during the summer when the plant grew quickly, and the dark bands in the winter when growth was slower.

Larger than life
tail of *Leedsichthys*

SIMPLE AND SMALL, the earliest forms of life were cells just a few thousandths of millimetres long (p. 9). In three and a half billion years life has evolved into a spectacular diversity that includes the massive 170-tonne (190-short ton) blue whale – the largest organism to have lived on Earth, ever. Life has increased in size by 16 orders of magnitude, but why?

Edward Drinker Cope (1840–1897) was encouraged by massive dinosaur bones unearthed in North America to propose that all lineages naturally tended to get bigger over time, a pattern that became known as 'Cope's Rule'. The evidence from the fossil record seems overwhelming. Take *Leedsichthys*, a truly titanic filter-feeding fish that reached a colossal size. Only isolated remains have been found like this magnificent 3-metre (9 feet 9 inches) tall tail from middle Jurassic rocks near Peterborough, England. A tail of this size could have come from a fish 17 metres (54 feet) in length.

It doesn't take a major evolutionary leap to get big. A tweak in the age at which adulthood is reached or faster growth will both make a larger body. And there are some very good reasons why bigger is better. Large size confers considerable advantage when hunting prey, and large size can help an organism survive times of shortage. Being big is also a sure way to reduce the chances of being eaten, and if there is competition for mates, size undoubtedly matters.

Although there are a plethora of examples that validate Cope's Rule there are also many counter-examples. One study showed that lineages appear to be just as likely to stay the same size as to get bigger over time. Who could argue that modern-day bacteria, single-celled algae and other microorganisms have not been successful, and they have had an exactly equal amount of time in which to evolve as the blue whale has. The trend towards bigger size may simply be a result of passive evolution. As organisms change and diversify through the vagaries of selection, lineages may increase or decrease in size. However, those that get bigger are more likely to survive because all organisms have a minimum functioning size. The net result is a general drift to an average larger size over time. If that is true, where are all the giants? Although being big brings its benefits, the fossil record tells us that it also brings an evolutionary risk at times of change. During the stress of mass extinctions, for example, the largest organisms tend to be hit the hardest. This periodic filtering may keep a cap on the tendency to get big, and may explain the loss of the world's largest creatures, like *Leedsichthys*, the ultimate big fish story.

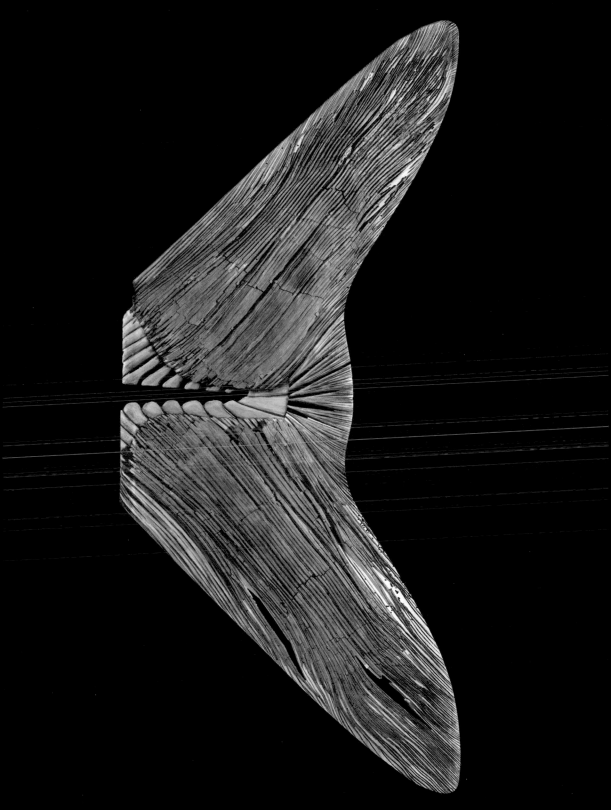

Correlation
Smith's ammonites

AMMONITES ARE PERHAPS the most iconic of all fossils. Related to modern squid, these cephalopod molluscs are found in great numbers in rocks of Jurassic and Cretaceous age. Their coiled shells form perfect spirals and are greatly sought after by fossil collectors. The Jurassic ammonite illustrated here is from the collection of William Smith (1769–1839), an English engineer who contributed so much to the young science of geology that he is often referred to as the 'father of English geology'. Illustrated by Smith in his 1816 book entitled *Strata Identified by Organized Fossils*, this particular ammonite was recovered from a well dug during the construction of the North Wiltshire Canal in southern England.

One year earlier, in 1815, Smith had published a remarkable geological map of England and Wales, the first of its kind, showing the bedrock geology that lies beneath the soil, each geological unit or stratum distinguished by its own colour. Smith was able to produce this map because he had worked out how to correlate units of layered sedimentary rock from place to place, from one outcrop to the next, across terrain where the bedrock is covered by soil and vegetation. This is no trivial matter, as rocks that look similar can come from different parts of the geological succession. Kimmeridge Clay of Jurassic age, for example, superficially resembles London Clay of Eocene age – both are grey in colour but London Clay is 100 million years younger than Kimmeridge Clay.

Smith's major insight was that the fossil content gave each unit a unique signature. Therefore, it was possible to correlate layered sedimentary rocks by their fossils, and conversely to use these same fossils to distinguish rocks of similar appearance but from different parts of the geological succession. For example, fossils present in the Kimmeridge Clay include ammonites like Smith's specimen of *Pictonia baylei*, a species that came to define a fine interval of geological time. Ammonites are absent in the London Clay, which instead contains crabs that would not look totally out of place in a rock pool today but are absent in the Kimmeridge Clay. We now know that such contrasts in the fossil content of sedimentary rocks of different ages can be explained by evolution and extinction. New species of animals and plants have appeared while others have become extinct, and this is duly reflected in the fossils present in sedimentary rocks formed at different times. Smith knew nothing of these principles that underpinned his recognition of strata by fossils because he worked during pre-Darwinian times, when fossils were generally attributed to the action of the biblical flood, the immensity of geological time was unappreciated, and evolution was not widely accepted.

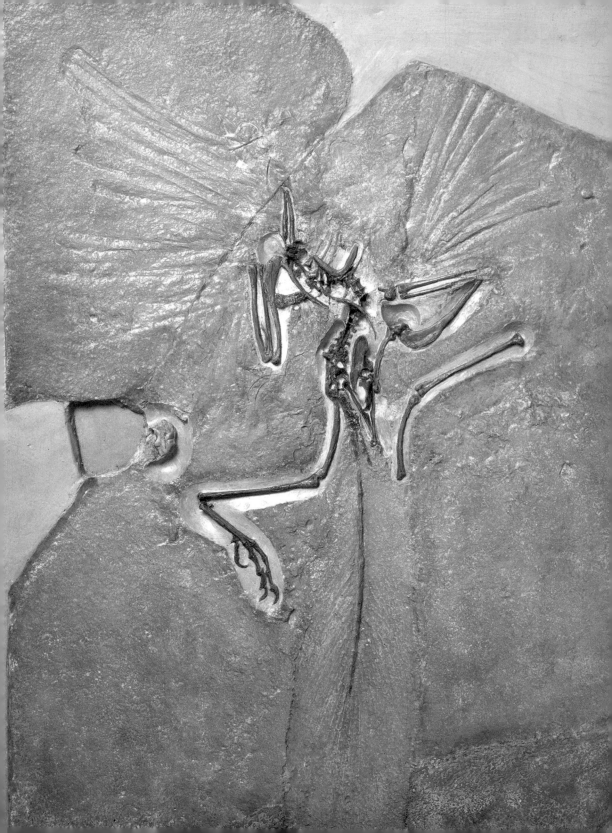

Birds hatch
Archaeopteryx

ARCHAEOPTERYX LITHOGRAPHICA IS ONE OF the most famous fossils of all time, widely hailed as the perfect evolutionary missing link. Known from only a dozen decent specimens, this 150-million-year-old animal is generally regarded as the oldest bird but one that possesses several reptilian features – notably teeth, a bony tail and claws on its hands. Unlike reptiles, however, *Archaeopteryx* has wings, feathers and a wishbone. The discovery of *Archaeopteryx* came only two years after publication of the first edition of *On the Origin of Species* (1859), providing just the kind of missing link that Charles Darwin (1809–1882) had sought to show – the continuity of groups of animals that differ so obviously today.

All examples of *Archaeopteryx lithographica* come from quarries around Solnhofen in Bavaria. Gemany. Here the Solnhofen Limestone was deposited as a lime mud in a quiet lagoon during the later part of the Jurassic period. Stagnant bottom waters meant that the carcasses of animals sinking to the bed of the lagoon were not disturbed or destroyed by bottom-dwelling scavengers and could become fossilized in a near-perfect condition when buried by lime mud. Feathers, for example, are preserved as delicate imprints in the fine-grained lithographic limestone.

Depicted here is the London specimen of *Archaeopteryx lithographica*, perhaps the most valuable specimen in the entire fossil collection of the Natural History Museum. It formed part of the fossil collection of a local doctor, Karl Häberlein (1828–1871), purchased by the British Museum. Häberlein had obtained the specimen in settlement of the medical bill of one of the quarrymen, and in turn sold it along with two thousand other fossils to provide a dowry for his daughter.

Although Darwin learned of the existence of *Archaeopteryx*, he scarcely mentioned it in his later works published after 1859. However, Darwin's vociferous supporter Thomas Huxley (1825–1895) appreciated the significance of this fossil, describing it in 1868 as the most reptile-like bird yet found. In the same publication he pinpointed dinosaurs as the most bird-like of all reptiles. We now know that *Archaeopteryx* evolved from a dinosaurian ancestor and that birds as a whole form a subgroup within a larger evolutionary group (clade) enclosed by dinosaurs.

Gigantic insects
dragonflies

INSECTS HAVE A VERY INCOMPLETE fossil record because they lack mineralized hard parts. Even so, an increasing number of fossil deposits are being found containing exceptionally preserved insects. These include not only amber but also fine-grained sediments that accumulated in shallow, tranquil freshwater lakes and marine lagoons. Among the latter is the famous Solnhofen Limestone of Bavaria, Germany, from which the magnificent dragonfly shown here was collected.

The Jurassic Solnhofen Limestone is a lithographic stone. Lithographic refers to the fact that this is a fissile stone able to be split along bedding planes with very flat surfaces ideal for applying ink for printing. The same bedding planes frequently expose the compressed remains of Jurassic animals. More than 50 species of dragonfly alone have been described from the Solnhofen Limestone, and this deposit is also the source of *Archaeopteryx* (p. 103) and numerous other exceptional fossils, including horseshoe crabs (p. 107), fishes and pterosaurs (p. 108). It is believed that both flying and aquatic animals sank to their death in the saline Solnhofen lagoon, the bottom waters of which lacked oxygen. Without oxygen, no animals could live on the bottom of the lagoon, and the limey mud that was accumulating in the lagoon quickly enveloped the corpses.

Dragonflies are primitive flying insects with two pairs of lacy wings. They are predators of smaller insects and usually live in wet environments, as their larvae are aquatic and develop in freshwater. The fossil record of dragonflies can be traced back to the Carboniferous, about 325 million years ago. Some Carboniferous and Permian griffenflies (the group from which today's dragonflies and damselflies are derived) grew to an enormous size: *Meganeura*, with a wingspan of 70 centimetres (27½ inches) would have dwarfed the largest dragonfly living today, which has a wingspan of a mere 16 centimetres (6.3 inches).

The existence of insects as large as *Meganeura* poses a major challenge. Insect respiration involves the passive diffusion of gases through a branching network of tubes called trachea. As insects become bigger, so the distance from the surface to the deepest parts of the tracheal system increases, making it more difficult for oxygen to reach these parts. In theory, this should limit the maximum size an insect can reach. Some scientists believe that the dragonflies of the Carboniferous could only grow to their gigantic size because of the higher atmospheric oxygen levels at the time – around 35% according to geochemical evidence. By contrast, today's atmosphere contains about 21% oxygen. With such a high concentration of oxygen in the atmosphere it may well have been possible for insects to grow much larger than they can today.

Living fossils
Mesolimulus

THE IDEA OF A 'LIVING FOSSIL' was introduced by Charles Darwin (1809–1882) in *On the Origin of Species* in 1859. He used the term to refer to two primitive animals – the duck-billed platypus and the South American lungfish. Both have endured to the present-day by inhabiting a confined area in which they are exposed to less severe competition. Exactly what constitutes a living fossil has been disputed, but still certain animals and plants are branded with the title. The most widely cited are the ginkgo tree, the coelacanth *Latimeria* (p. 126), and the horseshoe crab *Limulus*. They share the attributes of being the last surviving and relatively unchanged representatives of groups that in the distant geological past were more common, diverse and widespread. Living fossils are the tapering ends of formerly thick branches on the evolutionary tree of life.

This is the Jurassic fossil *Mesolimulus*, named for its strong similarity to the modern horseshoe crab *Limulus polyphemus*. The shield-like carapace has a pair of small compound eyes and covers the segmented part of the body bearing the limbs used in walking, swimming and respiration. A long tail projects from beneath the carapace, and in living animals acts as a lever to help animals on their back regain their correct orientation. These arthropods are not true crabs but are more closely related to spiders and scorpions. They belong to a group called the xiphosurans, which first appeared in the Ordovician period, nearly 480 million years ago; they were moderately diverse during the Palaeozoic, some inhabiting the sea but others living in freshwater or brackish environments. Like so many other animals and plants, xiphosurans were hit very hard by the mass extinction at the end of the Permian (p. 70). Indeed, just one lineage survived into the Mesozoic, and only four species of xiphosurans live today.

What evolutionary significance can be accorded to the living fossil *Limulus*? Unlike some living fossils, *Limulus* is widespread geographically, occurring along the Atlantic seaboard of North America and in Southeast Asia, and is very abundant where it does occur. Lacking a mineralized hard skeleton, the fossil record of xiphosurans is insufficient to know whether the group was ever represented by more species at any one moment in the geological past than the four that are living today. What is clear, however, is that xiphosuran morphology has changed little over time. Perhaps the horseshoe crab is an evolutionary example of the old idiom 'if it ain't broken, don't fix it' – an animal owing its extreme longevity to a 'design' that has been good enough to cope with the multifarious biological and physical changes that have occurred through more than 400 million years of Earth's history.

Powered flight
Rhamphorhynchus

OVER THE COURSE OF LIFE, several groups of animals have slipped the surly bonds of Earth and taken to the air to flee predators or to become apex predators themselves. Some manage only to jump and glide across short gaps whereas others may migrate enormous distances in search of food and procreation.

Gliding is an efficient way of moving through the air and is an approach employed by many small animals, such as squirrels, frogs and even ants. 'Flying' fish are in fact sophisticated gliders that use long pectoral fins to provide lift while their tail provides thrust. They can use this simple method to travel hundreds of metres, avoiding the jaws of a predatory fish or dolphin confined to the water. Some squid are able to do the same with limited lift from small tail fins but with the bonus of jet propulsion. However, while these fish and squid have managed a clever trick, true powered flight represents an evolutionary breakthrough pulled off by just a handful of groups during the last 350 million years. Today, birds and insects rule the day-lit skies while the bats carry the torch at night for mammals and with evolutionary aplomb. But long before the birds and the bats, pterosaurs were the first true flying vertebrates.

The fact that most pterosaur fossils are found in marine or lake sediments strongly suggests that fish were their principal source of food. There were continental pterosaurs that lived in forests and hunted insects, but these were small and better adapted to flying through forest than the open ocean. Living on land means they were much less likely to become fossilized, so their true diversity will probably be forever shrouded in mystery.

Meanwhile, the nautical pterosaurs are represented by a staggering 17 different families and hundreds of described species, one of which is represented by this fine example of *Rhamphorhynchus muensteri* collected from the Solnhofen Limestone in Germany. *Rhamphorhynchus*, or 'beak snout', was a long-tailed pterosaur. This particular fossil was the first one discovered in which the shape and structure of the wing membrane are preserved in great detail. The teeth of the upper and lower jaw show an intermeshing configuration, which is suitable for capturing slippery prey like fish.

Fishing they undoubtedly did, but they sometimes themselves became prey. In one remarkable slab of rock recently discovered the ultimate struggle between fish and pterosaur is preserved. A dead pterosaur lies prone on the ancient sea floor with a small fish still lodged in its throat but dwarfing both of these is a large, fast swimming predatory fish called

Aspidorhynchus with its sharp snout piercing the wing membrane of the pterosaur. Neat detective work by palaeontologists revealed the likely sequence of events. The unfortunate pterosaur had just swallowed the small fish, which it had caught by either diving into the water like a pelican or by flying low and dipping its snout into the sea. Almost immediately the large predatory and overly ambitious *Aspidorhynchus* lunged out of the water and attacked the pterosaur, found it too big to overcome and became entangled in its membranous wing. All three then fell to the sea floor, now beautifully preserved in fatal stop-motion.

Pterosaurs appeared 228 million years ago and survived until the end of the Cretaceous period 66 million years ago, where they joined the dinosaurs and ammonites in extinction. But for the previous 150 million years the pterosaurs ruled the skies as active flyers. And what a domination it must have been, with their often crested heads and taut wings silhouetted against the same sun that we look upon today.

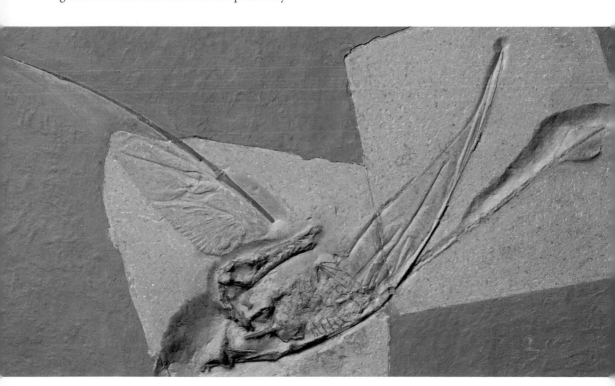

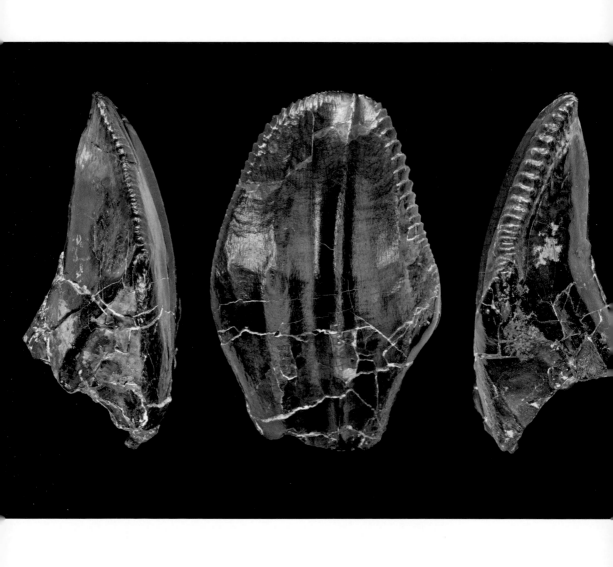

Dinosaur teeth
Iguanodon

A UNIQUE DINNER PARTY was held inside the mould made for a model of the dinosaur *Iguanodon* on New Year's Eve 1853. The model had been sculpted by British artist Benjamin Waterhouse Hawkins (1807–1894) for a display of prehistoric creatures that can still be seen in Crystal Palace Park, south London. Richard Owen (1804–1892), the leading British anatomist of the time, had instructed Hawkins on the reconstruction. This showed *Iguanodon* as a portly, almost dog-like animal supported on four thick straight legs, and with a long stout tail, small head and spiked nose matching the prevailing scientific view of dinosaurs as scaled-up lizards. Dinosaurs were envisaged by scientists as gigantic, slow-moving reptiles that lumbered lazily across the parched landscapes of the Mesozoic. How wrong they were.

This is a tooth of *Iguanodon*, one of the first dinosaurs to have been discovered. Mary Mantell (1799–1847), wife of medical doctor and fossil enthusiast Gideon Mantell (1790–1852), found the first specimens of fossilized *Iguanodon* teeth by a roadside in Sussex. Gideon went on to describe this new fossil as *Iguanodon* because of the similarity between the fossil teeth and those of the modern lizard *Iguana*, although the latter are far smaller in size. It was left to Richard Owen, who would later become a scientific adversary of Mantell's, to coin the name 'dinosaur' ('terrible lizard') in 1842 on the basis of *Iguanodon* and two other kinds of giant fossil reptiles. More and more fossils of *Iguanodon* began to be found in the Cretaceous rocks of southern England and subsequently in Belgium, where in one coal mine the skeletons of no fewer than 39 almost complete individuals were excavated between 1878 and 1881. Among these bones were the conical bones believed by Owen and others to be nose spikes.

Our view of dinosaurs, including *Iguanodon*, is now very different from that held in the nineteenth century. *Iguanodon* was a bipedal dinosaur with a large brain by reptilian standards. The question of whether all or just some dinosaurs were warm blooded, like mammals and birds, has yet to be settled, but few would doubt that these huge animals were anything but sluggish. As for those peculiar 'nose spikes', they are actually the thumb bones of the rather dexterous hand capable of grasping the plants on which this herbivore fed.

Webs of silk
spiders

IN 1726 JOHANNES BERINGER (1667–1740), a Würzburg scholar and physician, published a book that was to guarantee his notoriety in the history of palaeontology. Beringer's book depicted an amazing assortment of 'fossils' collected from a hillside close to Würzburg, Germany. Among the thousands discovered here by local people were spiders sitting astride perfectly preserved webs. Beringer used the preservation of the filigree webs as evidence that the spiders were not buried during the biblical flood, a popular explanation for the formation of fossils at the time, as the torrent would surely have destroyed them. It soon became clear to the hapless Beringer that he had been the victim of a massive fraud, probably perpetrated by his academic rivals – the spider and other specimens described by Beringer were carved by human hands and were not fossils.

However, spiders are not unknown in the fossil record. In recent years some remarkable fossils of spiders have come to light, including a few associated with silky threads. Spiders produce silk from spinneret glands on the abdomen, the silk emerging from modified appendages called spigots. The oldest known examples of such spigots have been found in a group of arachnids closely related to spiders from the Devonian, about 386 million years old. The original function of spider silk was probably to wrap the eggs in a protective covering, and the first spiders were apparently ground-dwelling predators. Only later did spiders evolve the ability to use their silk to spin aerial webs for catching flying prey.

Unlikely though it might seem, spider silk has been found fossilized after being enveloped by tree resin that became amber. Silk in amber dates back to the Early Cretaceous, with examples known from both the Lebanon and southern England. Even more extraordinary is this piece of Spanish amber, about 105 million years old, containing fragments of a spider's web that had entrapped a fly and a mite. Some scientists believe that the diversification of spiders themselves was linked to the evolutionary radiation of flying insects – the more flying insects, the greater were the feeding opportunities for web-constructing spiders.

Little did those responsible for the web of deceit that trapped Beringer realize that their fake fossils would eventually be more than matched by genuine fossils of spider webs almost as fantastical.

Great monkey puzzle
Araucaria

ENTHUSIASTICALLY IMPORTED BY Victorians and known to most as the monkey puzzle tree, *Araucaria* are highly symmetrical, very tall coniferous trees that offer a striking addition to parks and gardens around the world. In the wild, *Araucaria* trees are not particularly common, with natural populations restricted to isolated patches of mountain ridges or remote islands. The few species that have survived to the present day are so broadly distributed, from Hawaii and Australasia to South America, that they hint at a once wider distribution. Indeed the fossil record confirms this to be true. Fossil leaves, trunks and cones like the one pictured here, reveal that 150 million years ago, before the conquest of the flowering plants, much of North and South America, Europe and even Antarctica were blanketed by tall and diverse trees like modern *Araucaria*.

It has been suggested that the long neck of the sauropod dinosaurs evolved to reach the crowns of *Araucaria* trees, where more nutritious twigs and leaves could be browsed and digested slowly in their large stomachs. Millennia of this browsing in turn may have driven the evolution of ever taller trees and hence even longer necks, and so on, resulting in an evolutionary game of cat and mouse. Indeed, sauropod and araucarians both reached the pinnacle of their diversity and abundance simultaneously in the Late Jurassic and Early Cretaceous.

Walking through the *Araucaria* forests of South America today, one is drawn not only to the stunning canopy of the living trees but also to the ground strewn with the dead, fallen trunks and stumps. Once felled, the wood bakes in the dry air, taking on an eerie silver sheen, superficially resembling enormous dried bones. This resemblance led one palaeontologist to propose that these pseudo-skeletons gave sauropods the impression of a dinosaur graveyard, which kept the grazers out of the forest and saved the living trees from a munching. It's a tantalizing thought and, if true, a stroke of evolutionary genius on the part of the trees.

Sexual selection
Confuciusornis

DARWIN'S THEORY OF EVOLUTION by natural selection explains the preservation, over time, of traits that allow an organism to survive. Although the phrase was not used by Charles Darwin (1809–1882) himself, natural selection has come to be summarized as the 'survival of the fittest'. Sexual selection, by contrast, is the evolution of features that help maximize an organism's reproductive output by making themselves as attractive to the opposite sex as possible, best described as 'survival of the hottest'.

Looking around at the modern world it's clear that sexual selection has pushed some animals to extremes. Take the peacock, with its stunning, female-luring display of feathers. The drab, grey peahens are extremely fussy and only choose males with the most handsome train (for it is not a tail). The result is that a peacock needs a winning train if it is to successfully spread its genes, even though the train is costly to produce, is a hindrance when feeding and flying, and is a calling card to predators. Sexual selection appears to have overcome natural selection.

Despite being a major force in evolution, it is challenging to find robust examples of sexual selection in the fossil record. The horns of the herbivorous dinosaur *Triceratops* may well have been used in male-on-male battles for dominance over females, implying that their evolution occurred through sexual selection; yet equally the horns could have presented a formidable protection against predators, which would support the work of natural selection.

The best examples come from the kings and queens of sexual selection today – the birds and their reptilian allies the dinosaurs and pterosaurs. *Confuciusornis*, one of the oldest known birds with a beak, possessed a beautiful pair of long tail feathers that were probably used in sexual display. *Epidexipteryx*, an early, feathered dinosaur that was unable to fly, had four long tail feathers that would only have been useful for display and seduction. Some fossils of the pterosaur *Darwinopterus* have a head crest, but not all. Those with crests had thin hips, and those without had wide hips, suggesting only the males possessed the crests and that they were used as a mating display, like the bright-red throat pouches of the frigate bird. The hypothesis was confirmed by the find of a remarkable example of a female *Darwinopterus* preserved with egg, wide hips and no head crest.

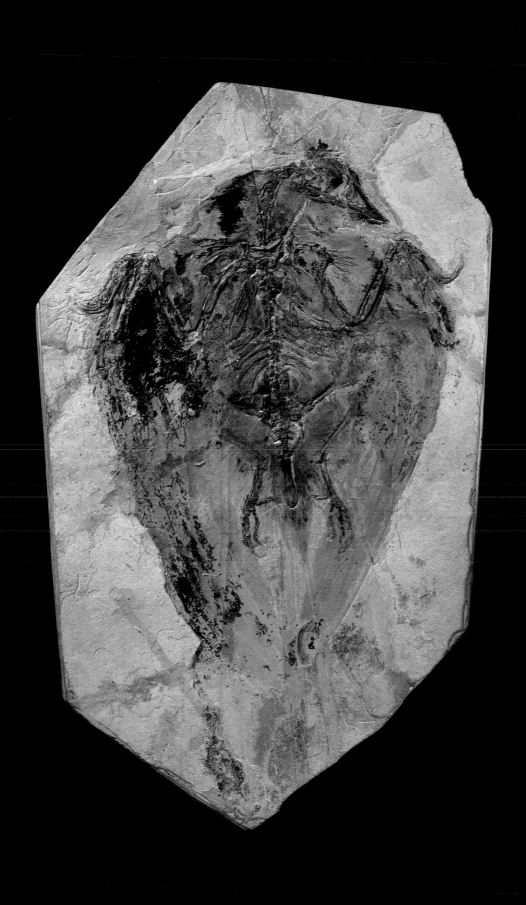

Losing a skeleton
octopus

OCTOPUSES ARE PECULIAR CREATURES. With their large eyes and head seemingly capable of accommodating an oversized brain, these molluscs resemble humanoid aliens from a science fiction film. The eight writhing, sucker-bearing arms, coupled with an ability to change colour to fit in with the background, add further to the 'other worldliness' of these animals.

As octopuses lack a mineralized skeleton, it is not surprising that they are uncommon in the fossil record. Some primitive octopuses do, however, possess a tough organic internal shell called a gladius, which is occasionally fossilized. But the octopus fossils that provide us with the most information are some exceptional examples preserving soft tissues. The specimen of *Palaeoctopus newboldi* shown here is just such a fossil. It was collected by the soldier and traveller Lieutenant Thomas Newbold (1807–1850), after whom the fossil was subsequently named, in the Lebanon around the middle of the nineteenth century. Not much more than the outline of the animal is visible, but this is sufficient to show that *Palaeoctopus* had eight arms. The arms have suckers, like all modern octopuses, but there are also indications of fins on the outside of the body, structures that are comparatively rare in octopuses living today. Unfortunately, a fossil fish is superimposed over part of the body of *Palaeoctopus,* slightly obscuring our view of the octopus.

Palaeoctopus was collected from rocks of Late Cretaceous age. Until quite recently it was the oldest octopus known in the fossil record. Examples of other fossil octopuses of roughly the same age (100 million years), or slightly older, have since been identified. There is even a disputed fossil octopus from Carboniferous rocks over 300 million years old, but this fossil has 10 tentacles and may not be a true octopus.

Octopus evolution bucks the common trend for marine animals to evolve thicker skeletons through time. Instead, octopuses show a progressive loss of the skeleton. This parallels a change in lifestyle, from squid-like ancestors that swam continuously at varying levels above the seabed, to fully fledged octopuses that spend much of their time on the seabed concealed in crevices, where the hard shell becomes a hindrance, and swimming only occasionally.

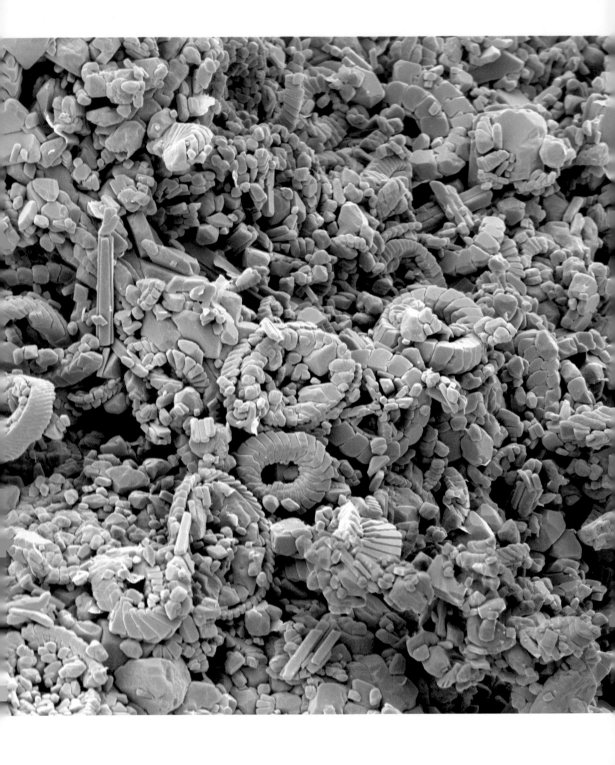

Chalk cliffs
coccoliths

THE FAMOUS WHITE CLIFFS of Dover on the south coast of England are a striking feature, visible when crossing the English Channel by sea. They were even mentioned in Julius Caesar's account of the Roman invasion of Britain in AD 54. But why are these cliffs so white? The reason is that they consist almost entirely of very fine grains of calcite, a mineral form of calcium carbonate, with little of the iron or other minerals that tend to give such layered sedimentary rocks their distinctive colours.

Chalk was deposited in the seas that covered a huge area from northwestern Europe to central Asia between about 100 and 66 million years ago, during the Late Cretaceous period. While dinosaurs were stalking the land surface, this ooze of fine sediment was accumulating on the seabed. The sediment is not mud eroded from the land and carried into the sea by rivers but instead is a biogenic deposit, consisting mainly of the skeletons of tiny algae called coccolithophores. Relatives of these planktonic plants float in the sea today, photosynthesizing and manufacturing exquisite skeletons made up of rings of submicroscopic plates called coccoliths, which protect the cells within. Coccolith blooms can occur following population explosions, turning the sea milky white in colour.

The coccolith-rich chalk of Cretaceous age was especially widespread because of the high sea level at that time, when areas of continent that are nowadays land were flooded by water up to a depth of 500 metres (550 yards). Coccoliths are so small that more than one million would be needed to cover the surface of a coin 2 centimetres (¾ inch) in diameter. Therefore, a very large number of coccoliths were required to form the great thicknesses of chalk – sometimes exceeding 200 metres (219 yards) – found in northern Europe. Even given the 34 million years of time it took the chalk to accumulate, this implies that the sea was extremely productive and very possibly had a milky white surface as is seen today during blooms of coccoliths. Up to 15 centimetres (6 inches) of coccolith ooze was deposited on the seabed every thousand years, which is at least three times the rate of deposition of coccoliths in modern oceans. The white cliffs of Dover and elsewhere are testimony to the remarkable flourishing of coccolithophore algae in the surface waters during the Late Cretaceous.

Pumping machines
chalk sponges

SPONGES ARE THE MOST primitive animals that have bodies composed of more than one cell. They have no nervous system, gut or any of the other organs we take for granted in animals. Today most sponges live on the sea floor, where they are highly efficient biological machines, pumping water through the pores and chambers in their bodies and filtering out bacteria and other tiny organic particles for food. Incessant movements of whip-like structures called flagella are used to generate these feeding currents, which end with the jet-like expulsion of filtered water from holes called oscula.

Molecular fossils – biomarkers (see p. 97) – of sponges have been discovered in rocks dating back almost 650 million years, but the body fossil record of sponges begins a little later, arguably in the Ediacaran rocks of South Australia. Here, the peculiar and distinctive assemblage of fossils (see p. 14) includes a genus that has been identified as a sponge on account of the spicules embedded in its body. Spicules are needle-like structures found in the majority of sponges. They can take many different forms from simple rods with pointed ends to formidable sharp rays joined at the centre, perfect for providing protection from grazers, and they may be composed of the protein spongin, calcium carbonate or opaline silica. Sponges with spicules of calcium carbonate or silica have a good chance of being fossilized, especially when the spicules interlock or are fused together to form a solid structure. This possibility is increased still further in sponges that also manufacture a robust calcareous skeleton. These include the stromatoporoid sponges that, alone or together with corals, constructed reefs in shallow waters during the Palaeozoic.

Some of the most remarkable fossil sponges, including the example of *Polyblastidium* shown here, come from the Late Cretaceous Chalk of Europe. Specimens etched from the chalk using dilute acid may reveal exquisite details of the spicules and the oscula through which the sponge expelled filtered seawater. Although sponges are common fossils in the Chalk, even more would be present were it not for the fact that sponges with spicules of opaline silica were vulnerable to dissolution soon after they became buried in the coccolith-rich mud making up the Chalk. However, an indication of their former presence can be found in the guise of flint nodules. Composed of silica, flint nodules are believed to have formed just beneath the seabed through the aggregation and eventual solidification of the silica dissolved from buried sponges. In a manner of speaking, every knobbly lump of flint is a fossil.

Arms race
Tylocidaris

HISTORY RECOUNTS NUMEROUS times when enemy countries or rival states have become trapped in an escalation of competitive military power. War is the anticipated product of these arms races, yet sometimes agreements are reached and treaties signed in the nick of time. There are arms races in evolution too that also involve a similar one-upmanship, but usually it is between predators and their prey. Nonetheless, the evolutionary result can be similarly extreme, with ever more sophisticated tools and tricks evolving in the predators countered by the evolution of equally innovative methods of defence, escape and avoidance on behalf of the prey.

The Mesozoic Marine Revolution was a time when arms races between predators and prey reached fever pitch. There was a rapid diversification of the number of different predators, with enhanced skills and innovative tools for catching prey. The prey that failed to evolve reciprocally to this treacherous new world went extinct, while the survivors were more heavily built, had greater fortification or moved deep into hiding.

Although the earliest echinoid, or sea urchin, appears in the Ordovician period, some 450 million years ago, it's not until the height of the Mesozoic Marine Revolution, some 200 million years later, that fossil sea urchins start to appear with telltale marks of them being predated. Sea urchins responded with the production of longer and thicker spines, like the ones seen in this astonishing specimen of *Tylocidaris* collected from the Cretaceous Chalk of southern England.

Urchin spines are strong, stiff and lightweight and, rather incredibly, each spine is formed from a single crystal of calcium carbonate. When a spine is lost, urchins can grow another easily and quickly. Each spine rotates at the base on a ball and socket joint. Muscles at the base enable the urchin to move each spine independently and in any direction, and also lock them into place, making them extremely useful in preventing a predator from gaining access to the urchin's nutritious central body.

The spines of *Tylocidaris* are club-shaped, solid and extremely robust, providing an almost impenetrable barrier to even the most determined fish, crab or other predator. They are proof of the lengths evolution can go to in an arms race. Rather than club-shaped spines, many modern-day sea urchins have sharp spines, some of which contain venom, are heavily barbed and difficult to remove from flesh. Next time you step on a sea urchin, instead of cursing the beast, grit your teeth and think of the wonder of the Mesozoic Marine Revolution.

Hooked from deep time
coelacanth

THE BIGGEST BREAKTHROUGHS in palaeontology don't necessarily come from the rocks, don't have to be fossils and aren't always made by palaeontologists. In 1938 a South African trawler captain named Hendrik Goosen (1904–1990) landed a large, strange blue fish. Goosen called the curator of the local East London Museum, an enthusiastic naturalist who often bought unusual fish from the docks for her collection. Marjorie Courtenay-Latimer (1907–2004) knew immediately she had stumbled upon a scientific curiosity, but unable to identify the specimen she posted a sketch to J L B Smith (1897–1968), a chemistry lecturer and fish biologist 50 miles away in Grahamstown. Even from the rough sketch Smith instantly recognized the large bony fish to be a living coelacanth – a member of a group of fishes that was until then known only from the fossil record spanning from 400 to 70 million years ago and presumed to have gone extinct along with the dinosaurs.

Setting eyes on the specimen Smith realized, with growing excitement, that his conclusion was indisputable. With its distinctive shape – Smith called the coelacanth 'Old Fourlegs' – and its multiple-lobed fins, the fish was indistinguishable in profile from fossil specimens such as this beautifully preserved specimen from the Cretaceous of England. The discovery

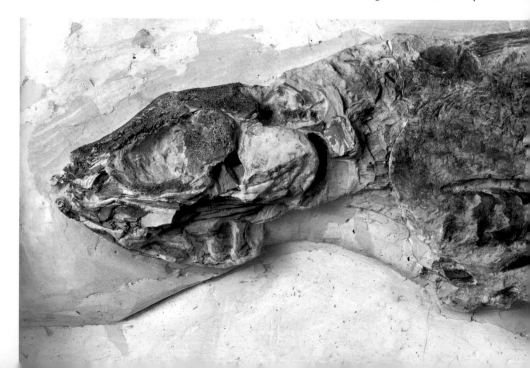

created a worldwide sensation and became the 'biological find of the century', revealing not only that coelacanths remained relatively unchanged for more than 400 million years, but illustrating starkly how little we know about life in the world's oceans today.

Desperate to find another specimen, Smith posted reward posters along the east coast of Africa and finally, 14 years later, a second coelacanth was captured off the Comoro Islands. So treasured was the specimen that Smith slept beside it during transport. Intensive searches since then have been decorated in controversy and disappointment, but finally several catches of coelacanths off Africa and more recently in Indonesia have provided great insights into this ancient lineage. Recent breathtaking observations of living coelacanths have revealed even more. Their large eyes are adapted to dim light ideal for their deep and secret habitat and their many fins make them extremely agile. Coelacanths give birth to live young, possess electrosensory organs to detect their prey, and are the only vertebrate to be able to open their mouth by hinges on both their lower and upper jaws.

The coelacanth's heyday was in the Devonian period; over 100 species have been described from fossils of that period found worldwide. Today two living species are known to science, and both are at risk of extirpation by overfishing as they potentially become caught up in the increasingly technological approach to extracting food from the sea – usually inadvertently as the coelacanth itself is considered too fatty and induces diarrhoea in humans when eaten. Protecting the coelacanths is an enormous challenge when so little is known about their biology and ecology. Perhaps there are additional species to be discovered on docks and market stalls in exotic locations, unwittingly hooked from the deep and from the past.

Parental care
nest of *Troodon*

THERE ARE PLACES IN THE WORLD where the remains of dinosaur eggs litter the ground like shards of ancient pottery. The scattered shell fragments are usually beyond reconstruction, and finding an intact egg is a rare event. A complete nest of unhatched eggs is therefore a highly prized find and one that commercial fossil hunters will go to great lengths to secure. Such rarities are also extremely valuable to the detective palaeontologist, revealing as they do glimpses into the private lives of dinosaurs.

All major dinosaur groups appear to have made nests. Some nests are more complex than others and, despite their rarity, many different types of egg clutch have been unearthed. This remarkable 77-million-year-old nest found in Montana was made by the parent dinosaur, almost certainly a female, by creating a hollow in the ground about a metre (3 feet 3 inches) in diameter with a raised rim to keep the eggs together. The ten eggs were laid vertically in the soft ground, pointing slightly inwards. The eggs themselves are teardrop-shaped, just like a chicken's egg, which helped to prevent them rolling out of the nest.

Large eggs must have come from large dinosaurs, but apart from that the palaeontologist really has little to go on, and struggles to identify the dinosaur species responsible. The eggs from Montana were initially thought to have been laid by a herbivorous dinosaur called *Orodromeus*, whose fragmentary skeleton was found near the nest. But scans of the eggs later revealed a remarkable set of embryos preserved inside the eggs – 'babies' of the small (1.8 metres/6 feet long) bird-like dinosaur *Troodon*. Other nests of *Troodon* show that this predator laid up to 24 eggs in a single nest. The arrangement of the eggs suggest they were laid in pairs, implying that *Troodon* had two functional oviducts, making its reproductive organs more like a crocodile than a bird, despite its overall similarity to birds.

Ensuring that eggs hatch and the youngsters survive is a battle of beating the odds. Turtles lay lots of eggs that, once deposited, receive little or no parental care. Crocodiles guard their nests but leave their offspring to their own devices shortly after hatching. Many such offspring are eaten by waiting predators before and after hatching, but the vast numbers laid mean that a few have a chance of making it to adulthood. Birds adopt a different approach and lay fewer but larger eggs, each one more likely to survive because of the high level of care given to it, sometimes by both parents.

The size of the eggs of *Troodon*, relative to the size of the adult, reveals that this dinosaur chose an approach somewhere between the many-eggs approach used by reptiles and the greater level of individual parental care that is invested by birds. The internal scans reveal stages of skeletal development in the embryos that would have required between 45 and 60 days of attendance by the adult between laying and hatching, but we simply don't know if the adults would have stayed around long enough to see their little brood become entirely independent. Either way, this nest provides evidence of an advanced level of parental care by *Troodon*, giving us an insight into the complex behaviour evident in one of the last dinosaurs to evolve.

Polar dinosaurs
Edmontosaurus

STEAMY SWAMPS, LUSH RAINFORESTS and scorched deserts provide the backdrop to many a classic dinosaur diorama and it is true that the highest diversity and greatest abundance of dinosaur remains are found in regions of the world that spanned the tropical belt. But in recent years a remarkable collection of dinosaur fossils have been unearthed from the North Slope of Alaska. What were they doing in the Arctic Circle?

Both herbivorous and carnivorous dinosaurs are present at the site, the commonest bones being from *Edmontosaurus*, a large duck-billed dinosaur that used its beak-like snout to browse on plants. Fossilized leaves reveal that for parts of the year the area was covered in a luxurious forest with abundant flowers, ferns and cycads. Clearly the world was much

warmer 70 million years ago than it is today, explaining why the plants (and the dinosaurs) didn't freeze, but it fails to explain how dinosaurs survived through the ever shortening days of winter that culminated in two months without sunlight.

One hypothesis proposes that the plants overwintered while the dinosaurs migrated south, much like reindeer do today. However, a lot of the bones are preserved in spring flood sediments, strongly suggesting that *Edmontosaurus* stayed put for the winter months. And if the herbivores like *Edmontosaurus* overwintered there would be no reason for the meat eaters to migrate.

So what did the herbivores like *Edmontosaurus* eat when the Arctic winter set in and plants stopped growing? Although hibernation has been suggested, a completely satisfactory answer remains to be found, but it does open an intriguing conundrum: if a wide variety of dinosaurs could survive such harsh conditions for months on end why did they all suddenly succumb to the 'nuclear winter' following the asteroid impact at the end of the Cretaceous? There is still a lot to learn about dinosaurs, how they became so widespread and so successful, and why their reign so suddenly came to an end.

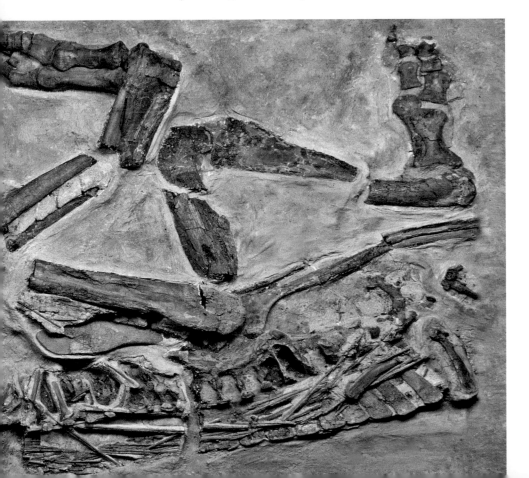

Grassed up
early grass phytolith

THERE ARE OVER 10,000 SPECIES of grass in the world today, and they can be found in almost all habitats, from floodplains to mountain ridges and on all continents, including Antarctica. What can the fossil record tell us about this extraordinary evolutionary triumph?

The earliest unequivocal fossil grass appears in the fossil record long after the dinosaurs went extinct 65 million years ago, so until recently most evolutionary biologists believed that it was only after the extinction of the dinosaurs that the grasses took root. Indeed, the diorama that placed dinosaurs against a backdrop of swathing grassland was swiftly lambasted by many an incredulous palaeontologist. However, in palaeontology, firmly held theories can be challenged by a single new find. Recently, a group of meticulous palaeontologists collected the coprolites (fossilized dung) of sauropod dinosaurs from India with the aim of reconstructing the diet of the massive herbivores. The team dissected all that was preserved within the coprolites and discovered a number of very small, glassy objects called phytoliths. Phytoliths, or 'plant stones' from the Greek, are microscopic grains of silica that many plants deposit in their tissues. They keep the plant rigid and protect it from grazing by fungi. Grass phytoliths have a very distinctive shape, and there, in the 70 million-year-old sauropod poo, were grass-like phytoliths including the one illustrated here. Grasses not only coexisted with the dinosaurs, they were in fact eaten by them.

Even more remarkably, the team found not one, but at least five very different types of grass phytolith. Such a level of diversity implies that at the time this particular sauropod munched away, a diverse range of different grasses had already become established. These discoveries suggest that the grasses must have a much deeper evolutionary history. Indeed, a recent find of what may be grass phytoliths in 100 million year old rocks could push the age of grasses back 45 million years earlier than previously thought.

If that is the case, where are the grass fossils for the missing 45 million years? The quantity of grass-like phytoliths discovered in the dinosaur dung was much lower than one would find in the dung of animals such as horses and cows that consume large quantities of grass. Indeed, the contents of the dino-dung revealed that palms, flowering trees and conifers made up the majority of the sauropod's diet. Grasses must therefore have been just a small part of the forest's flora, explaining why they have so far failed to appear in the fossil record.

Maastricht mosasaurs
Mosasaurus

WHEN NAPOLEON BONAPARTE HEARD that the skull of a monster animal no longer in existence was being flaunted in Flanders, he reportedly wanted it for France. The concerned owner tried to hide the precious fossil – to no avail. The reward for bringing the skull to Paris was several hundred cases of wine, sufficient to motivate French revolutionary troops to commandeer the specimen and transport it to Paris, where it is on display to this day in the Muséum national d'Histoire naturelle.

The skull was found in a limestone quarry near Maastricht, which today is in The Netherlands, and was the first discovery of a marine reptile called a mosasaur. Living in the warm seas at the end of the Cretaceous period, mosasaurs were large and efficient predators. The runt of the mosasaur family was a species that reached a piffling 3 metres (9 feet 10 inches) in length. Other species were considerably larger, and the largest was a 17 metres (55 feet 9 inches) giant armed with an array of strong conical teeth used for biting down upon its prey. Mosasaurs were robustly built and could deliver a powerful bite. The lower jaws were double-jointed as in many present-day snakes and helped the mosasaur to position the struggling prey for swallowing. Most mosasaurs – and we know more than 84 species – probably swam like eels, with an undulating body and tail. Some may have had a

tail fluke, providing more powerful propulsion under water. They preyed on fishes, diving seabirds and even on smaller members of their own kind.

After its dubious removal from Maastricht, the original mosasaur fossil arrived in Paris where the famous naturalist Georges Cuvier (1769–1832), fortunately on hand, was able to study it. With acute observation he proclaimed the beast to be a lizard – an impressive conclusion, as we now know the mosasaurs are closely related to monitor lizards such as the Komodo dragon. Cuvier did not support the new fangled theories that had begun to emerge among his peers. They proposed that the forces of evolution, or the transmutation of one form of life into another, explained the progress of life on Earth. He did, however, realize that fossils like the Maastricht mosasaur represented animals that were no longer alive, and to explain this he proposed that they had disappeared during a series of catastrophic events, the last of which was the biblical flood. Although the proposed mechanism was erroneous, the idea of extinction was revolutionary and entirely correct. It is estimated that 99.99% of all species that have ever existed are now extinct, and only a very small proportion of those are entombed in rocks in the same way that the Maastricht mosasaur is.

Cuvier went on to venture that there had been a time in the past when the world was dominated by now extinct reptiles, an idea that was shortly after vindicated by a rush of spectacular finds of plesiosaur, ichthyosaur and dinosaur remains in Britain in the early nineteenth century.

Clams and oil
rudist bivalves

THE MIDDLE EAST IS THE custodian of over two-thirds of the world's oil reserves, and this is thanks to a curious sequence of events and some well-placed fossil clams. Oil forms when ancient marine plants and bacteria die and become incorporated into sediments. Millennia of heating inside the Earth starts to change the organic material from these dead organisms, first into a waxy material called kerogen and then, with the perfect amount of heating and pressure, into hydrocarbons. Lighter than water, hydrocarbons migrate upwards through the rocks that entomb them. Normally they seep into the sea or soils and are lost forever. However, sometimes they can accumulate in porous rocks, trapped under an impervious cap, whereupon they accumulate and form a reservoir and finally an oil field. Only the right combination of rocks can form the enormous oil fields for which the Middle East is famous.

In the Middle and Late Cretaceous, some 100 million years ago, the Middle East and North Africa were covered by a wide tropical ocean called the Tethys, which stretched from the Mediterranean to the Indo-Pacific. The Tethys was so warm that corals were unable to survive, but a group of extinct bivalve molluscs (clams) called rudists were perfectly adapted to the heat. Rudists took over the role of reef builders and built magnificent reefs all around the coastal seas of the long-gone Tethys Ocean. The reefs that the rudists built over millennia, made of thick calcium carbonate and full of cavities, created the perfect rock to act as a reservoir for migrating oil. Where there are rudists, such as on the Karun Shelf of southwest Iran, offshore United Arab Emirates, Auila in eastern Libya and the vast oil fields of eastern Arabia, one finds some of the most valuable oil in the world.

Legendary king
Tyrannosaurus rex

LIFE IS A CONSTANT STRUGGLE. Different species constantly compete for limited food, individuals of the same species compete for mates and even members of the same family fight each other to survive. The outcome of these sorts of competitive interactions propels evolution forwards. And so it is that the legendary king of the dinosaurs stands as a figurehead for evolution.

It must have taken some intense competition to evolve into one of the largest predators to ever inhabit the land. *Tyrannosaurus rex* had forward-facing eyes, perfect for chasing prey. Bones of prey with stress fractures and tendon avulsions have been found that suggest that *T. rex* was a vigorous predator and not primarily a scavenger as has been suggested.

Hatchlings of *T. rex* grew rapidly to reach colossal size at adulthood, during which time their morphology shifted. Young tyrannosaurs were probably lithe predators, but

the adults became powerfully built animals that could produce a bite force just short of 6 tonnes (6½ short tons). The largest adults were therefore able to crush the bones of prey that the juveniles couldn't touch. This differentiation reduced the amount of competition between adults and juveniles of *T. rex*.

Tyrannosaurus rex first appeared around 70 million years ago and was all set to reign, when 66 million years ago life on Earth experienced a global mass extinction. The end-Cretaceous extinction not only saw the demise of all non-avian dinosaurs but also the loss of huge numbers of animal and plant species in the seas as well as on land. The cause of the event is still hotly debated, although mounting evidence points to an impact of a large comet or asteroid on the present-day Yucatan peninsula of Mexico that disrupted the whole biosphere.

As with all change there are winners and losers, and many consider that the loss of the non-avian dinosaurs was critical to the subsequent success of the surviving birds and mammals. With the dinosaurs absent the threat from highly proficient predators like *T. rex* was reduced and there were more opportunities in the world's ecosystems for the mammals and the birds to exploit – which they did so magnificently.

The K–Pg extinction
belemnites

SQUID-LIKE ANIMALS CALLED belemnites were prolific in the world's oceans during the Jurassic and Cretaceous periods, between 200 and 66 million years ago. These curious fossils are shaped like missiles, one end coming to a point and the other end containing a conical cavity. The largest of these so-called 'guards' are almost 50 centimetres (20 inches) long. Broken examples show the guard to consist of radiating crystals of the mineral calcite, giving an almost inorganic appearance.

Dubbed 'thunderbolts' in folklore, belemnites have been collected for countless centuries. However, their identity as cephalopod molluscs did not become known until German physician Balthasar Ehrhart (1700–1756) published a dissertation about belemnites in 1727. The main clue came from the preservation in a few examples of a chambered shell inserted into the conical cavity of the guard. This clearly resembles the chambered shell of the modern nautilus, although it is straight rather than coiled.

We now know a lot more about belemnites, thanks to the discovery of more complete individuals. Unlike most modern squid, they lacked suckers on their arms and instead had small hooks, doubtless for grasping their prey. Belemnites would have used jet propulsion to

propel themselves swiftly through the water, with the pointed end of the guard leading and the head and tentacles trailing behind. Floatation was achieved through the gas contained in the chambered shell, the heavy missile-shaped guard acting as a counterweight to head and tentacles.

The abundance of belemnites declined towards the end of the Cretaceous and they changed from being globally widespread to being restricted to high latitude seas in both the northern and southern hemispheres. None survived beyond the end of the Cretaceous – the species shown here is among the last. In this respect they shared the fate of many plants and animals, including the last dinosaurs, which became extinct during the famous K–Pg mass extinction. Theories abound about the cause of this momentous event. In recent years an asteroid impact has been the most popular, but some scientists believe that huge volcanic eruptions, which formed the lava flows of the Deccan Plateau of India, were at least in part responsible. Whatever the cause of the mass extinction, the end of the Cretaceous witnessed profound changes in life inhabiting the land and the sea.

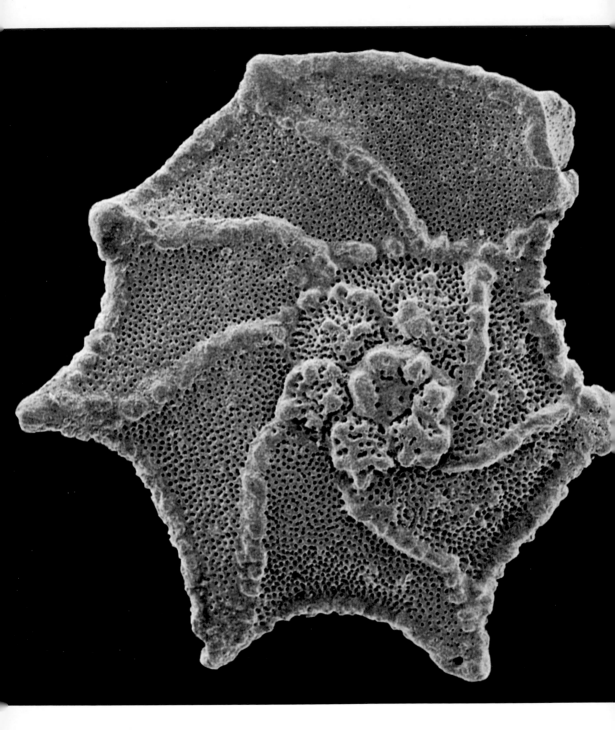

Small and simple
foraminifera

FROM SINGLE-CELLED BACTERIA to the intricate hovering flight of a hummingbird, life on Earth has repeatedly evolved towards ever-greater complexity. Such complexity clearly confers many advantages. Think of the defensive spines of the porcupine, the warning communications in tribes of baboons, and the fantastical mate-attracting displays of a peacock's tail. But complex isn't always the best option. The fossil record shows us that staying simple is sometimes a much safer bet.

Palaeontologists recognize two types of extinction; background and mass. Background extinction is the slow loss of species during times of little environmental change. Species that are widespread and abundant during background extinction are more likely to survive because having lots of individuals in different places acts as an insurance against unpredictable events. But during mass extinctions, when rapid environmental change wipes out large proportions of life over the entire globe, the dynamics of survival are different: being abundant and widespread doesn't always necessarily grant a species a safe passage. Instead, having a simple life style and a short life span is much more beneficial.

Take the foraminifera, a group of ocean-dwelling single-celled organisms that are abundant and important members of the floating plankton in oceans around the world. Forams, as they are affectionately named, create chalky skeletons that after death fall to the sea floor and form sediments known as oozes. The oozes that accumulate in the deep-sea are slowly laid down one on top of the other. By drilling through layers of oozes palaeontologists can observe the forams that lived both before and after the mass extinction that occurred 66 million years ago, at the boundary between the Cretaceous and Palaeogene periods. This most famous of mass extinctions saw off the dinosaurs and left a clear line in the rocks of the Earth.

Foraminifera survived, but before the K–Pg boundary they were extraordinarily diverse with hundreds of different species showing an array of shapes, sizes and ornamentations, a group clearly heading for increasing variety and complexity. After the K–Pg boundary though, the large complex foraminifera are missing – only some small and simple forms were able to survive. Some palaeontologists think that mass extinctions like the K–Pg, although rare, periodically curb the evolution of complexity. When it comes to mass extinctions it doesn't pay to be too big or too clever.

CENOZOIC ERA

THE FINAL ERA OF GEOLOGICAL TIME – the Cenozoic – stretches from about 66 million years ago until the present-day. Our planet and its biosphere really began to take their current forms during the Cenozoic. By the beginning of the Cenozoic all of the modern continents and oceans could be easily recognized, although with some important differences. North and South America were not joined together – the Isthmus of Panama linking them formed much later, about 3 million years ago. The Indian subcontinent was a large island in the Indian Ocean but drifted northwards to collide with Asia some 50 million years ago, a collision that caused the Himalayas to form. Australia was located in more southerly latitudes, closer to the Antarctic than it is today. The Tethys Ocean connected the Atlantic with the Indian oceans, forming an east–west seaway in the position of the modern Mediterranean Sea and across Arabia. Closure of the eastern end of the Tethys Ocean eventually left the now Mediterranean Sea with its only opening to the global oceans via the narrow Straits of Gibraltar. Just over 5 million years ago the Mediterranean became so isolated that it almost dried up altogether.

Accompanying these changes in the geography of the Earth were equally profound changes in its climate. Global temperatures peaked early in the Cenozoic, about 56 million years ago. Since that time the Earth has become cooler and has gone from a greenhouse to an icehouse climate. Cooling has not occurred evenly through time. Instead, there have been times when temperatures dropped sharply, including most recently when the Pleistocene Ice Ages began, about two-and-a-half million years ago.

The Cenozoic is known as the Age of Mammals for good reason – most major groups of mammals evolved rapidly during this era, and mammals became the dominant animals with backbones living on the land. Among these mammals were the primates, the group to which humankind belongs. Discoveries of new fossil primates are helping us to unravel the intricacies of the branches on the tree of life leading to modern man. Whales and dolphins, plus several other types of mammals, became secondarily adapted to life in the oceans, while bats took to the skies.

Birds are another group to have flourished in the Cenozoic, and both amphibians, including frogs, and reptiles continued to prosper and diversify. The dominance of angiosperms or flowering plants increased, with grasses appearing in abundance for the first time following their origin in the Mesozoic.

Among the invertebrates, insect diversity increased through coevolution with the angiosperms. Exquisite fossils of insects and also spiders were formed when trapped by oozing plant resin that hardened and became converted over geological time to amber. New groups of predatory snails appeared, many inhabiting reefs constructed by corals. In the oceans the familiar acorn barnacles first became numerous in rocky intertidal habitats, while crabs almost identical in appearance to those encountered in rockpools today made their debut.

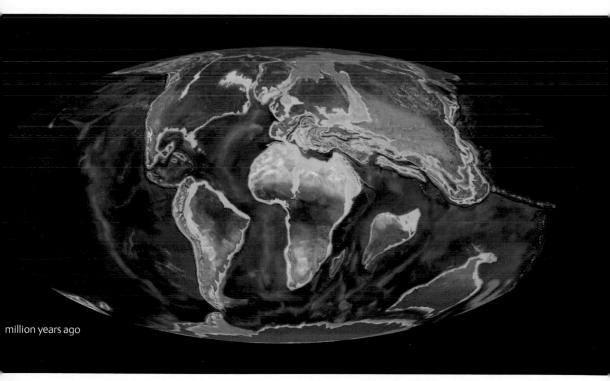

million years ago

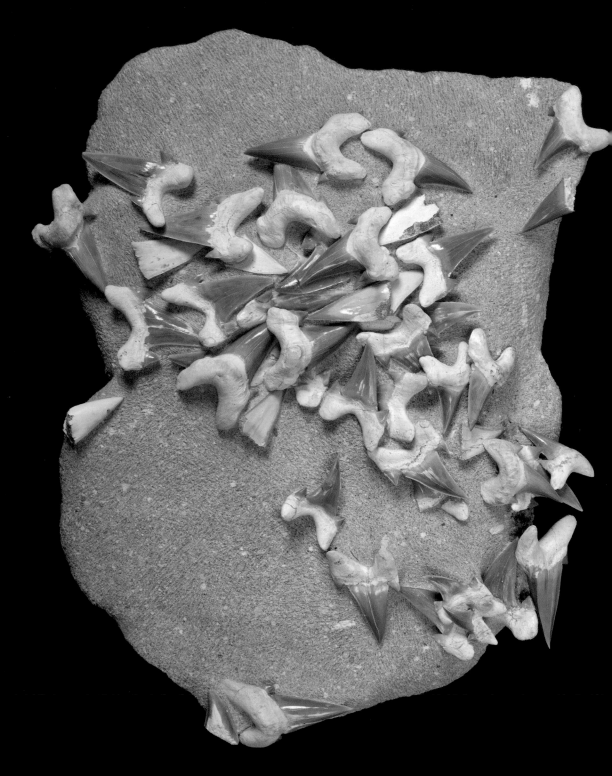

Formation of fossils
sharks' teeth

THE IDEA THAT FOSSILS ARE the buried remains of once living animals and plants only became widely accepted about 300 years ago. Before this time, the origin of fossils was a matter of considerable dispute. Many naturalists considered that a mysterious 'plastic force' caused fossil shells and bones to grow within rocks like minerals. Indeed, repeat visits to cliffs and pits seemed to show this growth actually taking place as particular fossils came to protrude more and more from the rock face with time. We now know that such observations are due to the progressive erosion of the soft rocks around the hard fossils, rather than growth of the fossils themselves. Even when fossils were correctly interpreted as the ancient remains of animals and plants, their specific identity and mode of formation caused difficulties for early naturalists. For instance, how could the shells of animals that lived in the sea be found on the tops of mountains hundreds of kilometres (miles) away from the coast?

One man is accredited with doing most to establish the true nature of fossils. Niels Stensen (1638–1686), or Steno as he is better known, was a Danish physician and naturalist who spent the latter part of his life in the service of Grand Duke Ferdinand II de' Medici in Florence. He was familiar with the curious fossils called 'glossopetrae' or 'tonguestones'. These had long been exported from Malta for their supposed medicinal powers. The story goes that in 1666 Steno dissected the head of a large shark landed at the port of Livorno on the coast of Tuscany. Noticing the close similarity between its teeth and glossopetrae, he came to the conclusion that the fossils must be the teeth of primeval sharks. Thus, these fossils at least represented the buried remains of animals that swam in ancient oceans at some earlier time, before dying and becoming entombed in sediment on the seabed.

Steno and contemporary naturalists would have possessed examples of the largest shark's teeth – those belonging to a species called *Otodus megalodon*, which lived between about 23 and 2.6 million years ago. Reaching 20 metres (65 feet 7 inches) in length, this huge animal, which has been found fossilized in rocks from many parts of the world, dwarfed even the modern great white shark. As with other sharks, new teeth grew constantly to replace older ones in a conveyor-belt-like manner. Each individual shark could produce a very large number of teeth with the potential of becoming fossils. Megalodon teeth embedded in fossilized bones demonstrate that this fearsome predator included large whales in its diet.

First flight
Onychonycteris

BATS ARE THE ONLY MAMMALS to have attained true powered flight. It's a demanding way of life that requires substantial amounts of energy and specialist anatomy, but once mastered evidently returns great evolutionary dividends (see p. 108). Today, bats are a remarkably diverse group, with over a thousand different species filling important roles within ecosystems around the world. There are two major groups, both of which are almost exclusively nocturnal. The microbats include the fish-eating, blood-lapping and insectivorous species, which all use echolocation to navigate and detect prey, whereas the megabats include the fruit-eating flying foxes, which don't use echolocation but have excellent eyesight.

Differences in the amount of mutation in the DNA of modern bat species pinpoint the common ancestor to the end of the age of the dinosaurs, in the Cretaceous period, yet despite extensive searches the oldest bat fossils come from rocks many millions of years after the dinosaurs' reign. This is a pattern that is repeated across many mammal groups, suggesting that although mammals arose in the shadow of the dinosaurs, they were unable to successfully expand until after the reptiles' departure. Once the bats became successful their chance of fossilization increased – it's a simple case of increasing the odds.

DNA evidence also suggests that all bats are monophyletic; that is, they have a single common ancestor. It follows that because all bats fly, it is likely that this common ancestor was itself capable of flight. The discovery of this stunning complete skeleton, of a bat named *Onychonycteris finneyi* in 52-million-year-old lake sediments from Wyoming, USA strongly support this. Clearly this bat could fly, as it has fully formed wings even while retaining the long tail and clawed fingers of its presumed tree-scrambling ancestor, which glided from tree to tree. However, the ear bones or cochlear in this, the most primitive bat yet found, are too small and the throat not sufficiently well developed to have enabled them to use sophisticated echolocation. It therefore appears that bats flew before they had evolved echolocation.

The question that now begs to be answered is 'Was *Onychonycteris* nocturnal or diurnal?' The eye sockets are unfortunately too poorly preserved to determine if, like the flying foxes, it could fly through twilight without the use of echolocation. That particular question will have to wait until even more remarkable specimens are unearthed that can shed light on the fascinating early evolution of the bats.

Back to the water
Rodhocetus

AS CHARLES DARWIN (1809–1882) correctly observed, reading the history of life from the fossil record is like reading a book that has lost some important character descriptions, is poorly illustrated and is missing whole key chapters, particularly early on in the story. But sometimes we get lucky and find a sequence of events that unveils evolution in action.

The move from water onto land by our early fishy ancestors must have been driven by the availability of new habitats and sources of food, and perhaps because of the protection land gave from increasingly efficient aquatic predators. Once on land, however, life became equally competitive, and it wasn't long before the water again became more favourable for some. And so it was that the dolphin-like ichthyosaurs, long-necked plesiosaurs and ferocious mosasaurs independently evolved from terrestrial reptiles and plunged back into rivers, lakes and seas.

Darwin realized that this evolutionary return to water must have taken place and went so far as to speculate, after observing the behaviour of a bear catching flies in water 'for hours', that if given enough time natural selection could lead to one lineage of the bears also evolving into an aquatic animal. The idea seemed absurd at the time, and Darwin didn't have the powerful evidence from the fossil record that we now have at our disposal, much less DNA, so he had no way of demonstrating its possibility. Since Darwin's time the puzzle has been greatly illuminated, though not yet completely solved, by the sequential find of amazing early whale fossils.

The ancestor of all the cetaceans (dolphins, whales and porpoises) was an even-toed ungulate – a hoofed animal that rests its weight on two toes, like pigs, deer, cows and hippos.

But whales, with their streamlined body, missing hind limbs and fluked tail, share little with their ruminant cousins that opted to remain dry. Most importantly, their forelimbs are so highly modified that they have lost the defining features that link them with their ancestors. Such a transformation seems at first glance as bizarre as the one Darwin proposed.

However, the fossil record narrates the return to the waters taken by the cetaceans. *Indohyus* was a raccoon-sized even-toed ungulate that waded in water 50 million years ago. Its bones were dense to help it sink and it had an unusually thickened inner ear bone, which aids hearing underwater, a feature that only cetaceans now possess. A few million years later *Rodhocetus* appeared. *Rodhocetus* was even more adapted to life in water than *Indohyus*, exclusively seen in even-toed ungulates, as illustrated here in this complete hind foot. Fossils show it had a short, thick neck, more paddle-like feet to efficiently move through water, a large tail to act as a rudder and even more efficient underwater hearing. Its legs were disconnected from the pelvis, so walking on land could not have been a comfortable or efficient experience, and yet *Rodhocetus* retained small hooves at the ends of its toes and an unusual 'double-pulleyed' ankle bone exclusively seen in even-toed ungulates, as illustrated here. After *Rodhocetus* there came *Basilosaurus* (p. 162), a much larger beast that was fully aquatic, with highly modified flippers, long tail and flexible body – yet it too retained vestigial legs that betray its ungulate ancestry.

These and other wonderful fossils piece together how the ancestor of a hippo took an evolutionary deviation and returned to the sea, radiating into all the species of whale and dolphin that we now know. Darwin's speculation on the fishing bear does not seem quite so ridiculous now.

Zombie death grip
Ophiocordyceps

THE EYEBALL-SUCKING SEA LOUSE, the five-mouthed sinus-feeding crab, and the copepod anal skin-bag are some of the more macabre examples of the innumerable and omnipresent parasites. Take comfort that parasites rarely kill their hosts, for to do so would threaten the number of hosts available to the parasite's descendants. In some distorted, roundabout way, parasites try to adhere to the old adage 'don't bite the hand that feeds you'. If a parasite does kill its host, it is called a parasitoid; death usually comes after no further benefit can be garnered from the host, and the loss of the individual does not greatly affect the survival of the host species.

One of the most gruesome examples of parasitoids that take control of their hosts, in a process termed 'zombification', is a fungal parasitoid called *Ophiocordyceps*, which infects worker ants foraging on the forest floor. Once infected, a bizarre sequence of events is set in motion. First the fungus penetrates the ant's body and spreads throughout it like a yeast. It takes hold in the ant's brain, filling the head, and begins to alter the ant's behaviour. At this point, the ant can no longer be considered an ant but more an ant's body controlled by the fungus within. Remarkably, the ant's body continues to move but in convulsive uncoordinated movements – a real-life zombie. Finally, the decisive moment arrives, when the fungus coerces the ant's body to climb upwards, away from the forest floor and onto the underside of a leaf, whereupon it reaches one of the veins of the leaf and with its mandibles clamps down hard around it – a behaviour unknown in healthy ants. After a couple of days in this 'death-grip' around the leaf's vein, the fungus begins to change yet again. It sprouts a stalk from the ant's head that is almost twice as long as the ant itself. The stalk is loaded with the spores of the fungus, which are then in a perfect position to spread themselves far and wide across the forest floor, ready to attach themselves to another unsuspecting ant.

The slicing marks left by the ant's death-grip on a leaf are unlike any other. After an exhaustive search of thousands of fossil leaves, palaeontologists discovered this single, 47-million-year-old laurel leaf from the Messel Formation, Germany, which preserves in beautiful clarity 29 telltale death-grip scars. Even without fossils of the fungus or the ant, these distinctive scars demonstrate that the fungus–ant relationship must have evolved a very long time ago. The ants that are parasitized are sterile, unable to reproduce, so their death may have only a small impact on the survival of the ant colony. The fungus can kill its individual host while ensuring the survival of the host species, and in this way the ants and the fungus have continued to survive for tens of millions of years.

Crabs and concretions
Chaceon

WHETHER SCURRYING ACROSS a tropical beach before descending into a burrow, or lurking with intent at the bottom of a temperate rock pool, crabs have a charm outweighing the threat of their claws. Crabs are crustaceans that first evolved from a lobster-like ancestor during the Jurassic period. The transition from lobster to crab involved reduction in length and bending of the tail so that it became tucked beneath the formidably armoured carapace – protecting this, the meatiest part of the animal, from would-be predators. In most crabs, including edible and shore crabs, the carapace also evolved a broader and flatter shape.

The earliest Jurassic crabs are uncommon, and not until the Eocene did crabs appear in significant numbers in the fossil record. This magnificent example of an Eocene fossil crab, *Chaceon peruvianus*, comes from Patagonia.

Finding fossil crabs as well preserved as this specimen of *Chaceon* is not easy – most fossil crabs consist of claws alone. Scavenging animals and currents quickly destroy the dead carcasses of crabs on the seabed, the skeletons becoming disarticulated and degraded. Experimental studies have shown that the legs are first to go among the hard parts, followed by the carapace, and lastly the claws, which are reinforced by thicker calcification to make them strong for their function. The most complete fossil crabs tend to be found in concretions. In this case, rapid burial of the crab carcass before disarticulation is followed by the growth of the concretion around the crab. Phosphate released from the rotting crab carcass, as well as other chemical changes in the vicinity of the crab, favours the precipitation of minerals between the grains of sediment, cementing the sediment and causing the concretion to grow. Encased within

its hard concretion, the crab skeleton is now protected from crushing and has a much greater probability of surviving intact as a fossil.

Whereas this Eocene species of *Chaceon* was collected from sedimentary rocks deposited in a shallow-water environment, species of the genus living today inhabit the deep sea. They include *Chaceon fenneri*, the commercially exploited 'golden crab' found in the tropical Atlantic. The fossil record provides evidence that numerous marine groups originated in shallow waters close to the shoreline, before embarking on a slow migration offshore into deeper waters through geological time. Some scientists have interpreted this to mean that the deep sea provides a refuge for animals driven out of more favourable shallow-water habitats by newly evolved, superior competitors.

Amoebae and pyramids
nummulites

WHEN THE GREEK GEOGRAPHER Strabo (63 BC to around AD 24) visited the Pyramids of Giza in Egypt, these wonders of the ancient world were already 2,500 years old. Strabo was duly impressed by the huge size of the pyramids, but also remarked on the lentil-like structures contained in the stones from which the pyramids were constructed. He was told that these were remnants of the workers' food turned to stone – an origin Strabo wisely regarded as improbable. We now know that the 'slaves' lentils' are fossils belonging to animals called nummulites that flourished in a warm shallow sea covering this part of Egypt about 40 million years ago. The name 'nummulite' alludes to the fact that larger specimens strongly resemble coins. Indeed, in folklore in Egypt they are referred to as 'angels' money'.

The simple single-celled structure of nummulites contrasts with their intricate skeleton, forming a series of spiral, overlapping whorls, with each whorl being subdivided into numerous tiny chambers. Nummulites may reach 10 centimetres (4 inches) in diameter, but, remarkably, they are the fossils of single-celled animals closely related to amoebae. Why did they grow so large? There is good reason to believe that, like their modern relatives, the nummulites found at Giza were hosts in a symbiotic relationship with another, smaller organism. In the case of species living today, these symbionts are tiny singled-celled algae called diatoms, golden brown in colour. The shells of nummulites are relatively transparent, and their flat shape provides a large surface area for the light required by the diatoms to photosynthesize effectively. For reasons that are still being debated by scientists, the presence of plant symbionts in animals living in the sea promotes the growth of calcareous skeletons in the host animals. Thus, the gigantic size of the nummulites at Giza was probably due to the intimate relationship they developed with symbiotic diatoms.

Their great utility in dating rocks, especially across the oil-rich states of the Middle East, means that nummulites have been intensively studied by palaeontologists. Species of nummulite evolved rapidly, and consequently their fossils change from one layer of sediments to the next. Furthermore, nummulite fossils are often present in prodigious quantities. The limestone used for the pyramids is so packed with them that, as at Giza, it is known as a nummulitic limestone. One of the fascinating facts about such nummulitic limestones is that they very often contain nummulites of two different sizes – the smaller 'slaves' lentils' and the larger 'angels' money'. These are not different species but rather distinct stages in the life cycle of a single species.

Caribbean cold seep
vent bivalves

A MOMENTOUS DISCOVERY was made in 1977 by a team exploring the deep sea floor in a region of the East Pacific called the Galápagos Rift. Here, a community of organisms was found clustered around a hydrothermal vent where extremely hot saline water gushed from fissures in the rock. Not only did this community contain many previously unknown species of animals, but it also represented the first example of a living community powered by geothermal energy rather than sunlight. Similar hydrothermal vent communities have since been located along other mid-oceanic ridges around the world. Since 1983, discoveries have also been made of a second type of seep community around vents issuing brines scarcely warmer than normal seawater. Like hydrothermal vents, these cold seeps contain bacteria that are able to metabolize the methane and hydrogen sulphide emanating from the vents. Larger organisms, including tubeworms and molluscs, either consume these chemosynthetic bacteria for food, or receive their nutrition from similar bacteria living symbiotically in their own tissues.

Several scientists have considered the idea that life on Earth may have originated at either hydrothermal or cold vents. But can we distinguish vent communities in the geological past? In fact, fossil examples of both hydrothermal and cold vent communities have been recognized in rocks dating back at least to the Silurian period. Fossil hot vent communities have been discovered in association with commercial deposits of sulphide ore minerals. Cold seep communities can be harder to distinguish, but some isolated bodies of limestone surrounded by muddy sediments show the chemical fingerprint characteristic of cold vents. One such example occurs on the Caribbean island of Barbados.

This rather unprepossessing fossil is a bivalve mollusc collected in Barbados from an Eocene cold vent community roughly 50 million years old. Very similar vesicomyid bivalves live today at cold vents in the sea around Barbados and elsewhere in the world's oceans. Living examples have poorly developed guts and rely for nutrition on the symbiotic chemosynthetic bacteria in their gills.

The initial expectation was that vent communities would be particularly ancient, providing scientists with a unique 'glimpse of antiquity'. However, research on fossil vent communities has shown that this is not the case: many groups of animals, such as vesicomyid bivalves, in modern vent communities are geologically recent arrivals, and older vent communities contain very different animals.

A Darwinian snail
Trophon

CHARLES DARWIN (1809–1882) returned on 2 October 1836 from almost five years of exploration aboard HMS *Beagle*. The seeds of a theory of evolution by natural selection were firmly planted in his mind and he had accumulated enormous collections that were to occupy him and his fellow naturalists for many years to come. Among the fossils was the Oligocene gastropod from Patagonia shown opposite. This was initially named *Fusus patagonicus* by the great mollusc specialist G B Sowerby (1788–1854) in 1846, but later renamed *Trophon sowerbyi*.

Darwin collected the fossil at Port San Julián in southern Argentina. According to his account of the voyage of the *Beagle*, the ship docked on the 9th January 1833, allowing him to explore the natural history in the vicinity. Here he located sedimentary rocks rich in fossil shells. As these shells clearly belonged to marine animals, Darwin deduced that this part of Patagonia must have been uplifted by at least 100 metres (328 feet), and that this occurred quite recently because, although belonging to extinct species, the shells were not greatly dissimilar from contemporary species living in the sea.

Darwin praised the work of his friend, geologist Charles Lyell (1797–1875) who was a strong proponent of 'uniformitarianism', which proposes that the slow processes that can be observed today, when allowed to act over geological time, can explain all ancient geological features. Lyell's teachings allowed Darwin to appreciate the immensity of geological time and also the dynamism of the Earth since its formation. The concept of an extremely old Earth became important to Darwin in providing sufficient time for the gradual evolution of all species from a single common ancestor. The changing environments of a dynamic Earth also ensured that there were always new evolutionary opportunities for living plants and animals.

A little before Darwin visited Patagonia, French naturalist Alcide d'Orbigny (1802–1857) had been there on a similar expedition. However, unlike Darwin who had been inspired by uniformitarian principles, d'Orbigny was a follower of Georges Cuvier (1769–1832), a 'catastrophist' who believed that many geological features can only be explained by catastrophic events no longer seen today. These catastrophes periodically brought about the total extinction of life on Earth, with a wave of newly created species subsequently repopulating the planet. With such contrasting influences, Darwin and d'Orbigny developed very different views about the geology and fossils of Patagonia, and of the history of life on Earth. While Darwin returned to England to develop his evolutionary ideas, d'Orbigny went back to France and expounded the catastrophic ideas of Cuvier.

A whale of a time
Basilosaurus

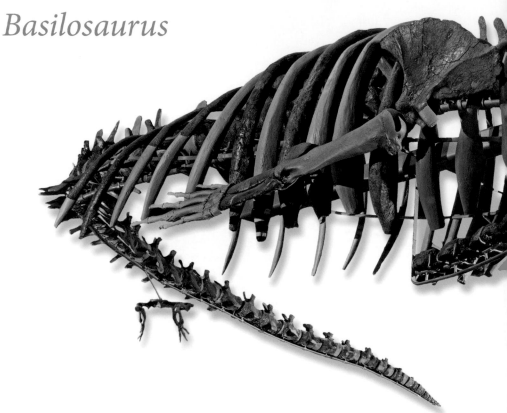

SOMETIMES RECONSTRUCTING extinct organisms from a fragmentary fossil record takes time. The early whale *Basilosaurus* took over 150 years to finally be complete. This large whale is now known to have reached up to 16 metres (52 feet 5 inches) in length when alive. It was first described in 1834 by American anatomist Richard Harlan (1796–1843) based upon a single massive vertebra and other fragments found in Alabama. Harlan coined the name *Basilosaurus*, which means 'king lizard', because he thought it belonged to an extinct carnivorous reptile. The natural historian and founder of the British Museum (Natural History) Richard Owen (1804–1892) redescribed the bone and new ones since found and in doing so recognized that the animal was in fact a whale.

However, a few years later, the fossil collector Albert Koch (1804–1867) made a new and large collection of *Basilosaurus* bones that included a skull, but failed to account for Owen's redescription and maintained they were from a 'giant sea serpent', which he called

Hydrarchos. Koch took many bones from different individuals to build up an impossible 35-metre (114-foot) long monster that he claimed would have weighed over 2.7 tonnes (3 short tons) and took the bones around the USA and Europe on a travelling, money-making exhibition. Fifty years later Smithsonian Curator George Brown Goode (1851–1896) managed to piece together a new hoard of bones to make a near-complete skeleton in 1894, but a number of tail bones were missing and so were added from another specimen and displayed for the next 100 years in the new Museum of Natural History in Washington DC.

During this time, however, the skeleton lacked its small, disarticulated hind limbs. It had long been known that *Basilosaurus* had hind limbs, and bones had originally been described as the hips of *Basilosaurus*. However, the identity of these bones was subsequently hijacked by Austrian palaeontologist Othenio Abel (1875–1946) who, working only from drawings of the hip bones, was convinced that they were in fact the shoulder bones of a long-extinct flightless giant bird. Abel published his hypothesis of the non-existent bird in 1906, adding unnecessary confusion to *Basilosaurus* evolution.

It took excavations by American palaeontologist Phillip Gingerich (1946–) to complete the puzzle. Gingerich described a remarkable set of *Basilosaurus* skeletons from Egypt, in which not only vertebrae, pelvises and hips were preserved, but fully functional 12 centimetre (4¾ inch) long feet with ankles and toes, demonstrating the intermediate position of *Basilosaurus* between older land mammals and the modern limbless whales of today.

Now, the famous *Basilosaurus* skeleton hangs in the Sant Ocean Hall in the National Museum of Natural History in Washington DC, replete with a full set of bones, bird hips and feet included, for all to marvel at.

Stuck in time
cockroach

IN SEVERAL RESPECTS, insects can lay claim to being the most successful group of animals ever to have evolved. Well over half of animal species living today are insects, and these invertebrates have prospered for more than 300 million years in almost all environments on Earth apart from the sea. Love them or loathe them, and most people fall unreservedly into the latter category, it is difficult not to be impressed by the sheer abundance and tenacity of insects. Indeed, one type of insect – the cockroach – has been portrayed as the world's oldest pest and the only animal likely to survive a nuclear holocaust. Cockroaches have a long geological history stretching back to the early Cretaceous, over 100 million years ago. But the example shown here encapsulated in Baltic amber is much younger, about 35 million years old. Cockroaches are large insects with a small head and flattened body, enabling them to shelter in crevices. The first pair of wings is modified to form a tough protective layer over the top of the broad body. They are very resilient animals, capable of surviving for long periods without food and even without air for up to 45 minutes.

Insects preserved in amber are among the most remarkable of all fossils. Amber is formed by the fossilization of resin that oozed out of particular types of trees. It was not uncommon for small animals, especially insects but also lizards, to become trapped in the sticky resin. Through geological time the resin hardened as heat and pressure drove off the more volatile components, to leave the semi-precious stone that has been used in jewellery for more than 10,000 years. Spectacular examples are known of beetles, flies, mosquitoes, gnats, bees, ants and many other types of insect in amber. Fuelled by the movie *Jurassic Park*, in which the DNA of dinosaurs is extracted from blood-sucking mosquitoes in amber, much media and scientific interest has been focused on the possibility of extracting DNA from amber insects. However, despite the exquisite preservation of anatomical features such as muscle fibres, cell nuclei and even mitochondria, DNA has yet to be extracted successfully from any insects preserved in amber.

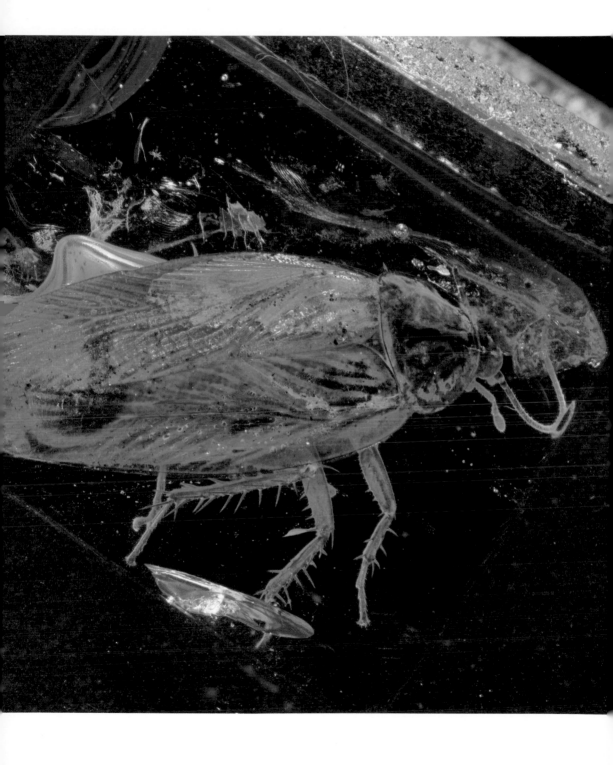

Origins of primates
Aegyptopithecus

THIS IS THE CAST OF A SKULL of the fossil mammal *Aegyptopithecus*. Along with humans, *Aegyptopithecus* belongs to an order of mammals called primates, which can be traced back in the fossil record about 60 million years to the Paleocene. *Aegyptopithecus* is a relatively early example of a primate that lived about 33–35 million years ago during the Oligocene epoch. Primates consist not only of *Aegyptopithecus* and humans, but also lemurs, lorises, tarsiers, monkeys and apes. They are a remarkably uniform group anatomically yet are highly specialized in terms of their behaviours. For instance, all primates have the same number of bones in their hands, and these have similar proportions; however, the way tarsiers use their hands differs greatly from how humans use theirs. Most primates are arboreal (tree-dwelling) animals inhabiting tropical forests, humans being an obvious exception to this rule.

As the name suggests, *Aegyptopithecus* was originally found in Egypt. It was discovered in 1965 in the Fayum Depression to the west of Cairo. Fayum is the palaeontological equivalent of the Valley of the Kings. It has been a bonanza for Egyptian vertebrate fossils since the first expedition there in 1879 led by the botanist and explorer Georg August Schweinfurth (1836–1925). In addition to apes, fossils of primitive whales (p. 162), sea cows, marsupials, animals resembling elephants/rhinos and hippos, and carnivorous, hyena-like creodonts have all been found at Fayum.

Aside from its antiquity as a primate, the importance of *Aegyptopithecus* is that it belongs to the ancestral group that gave rise on the one hand to Old World monkeys and on the other to apes. The most obvious difference between monkeys and apes living today is that monkeys usually have a tail whereas apes always lack a tail. In some reconstructions *Aegyptopithecus* is depicted as having a tail, whereas in others it is shown tailless. Although the brain of *Aegyptopithecus* was small – about the size of a lime fruit – the part of the brain responsible for vision was large, suggesting that the animal had good vision and was active during the day. Studies of patterns of microwear on the teeth of *Aegyptopithecus* provide evidence of its diet and point to occasional consumption of hard food such as nuts to supplement a staple diet largely of fruits.

Blooming plants
Chaneya

ONE GOOD WAY TO FOSTER evolution is to have sex, thus ensuring varied and changing offspring and increasing the chances of beneficial mutations. But how do you do it when you are rooted down, unable to move? In the sea it can be as simple as casting spores, sperm or eggs in the hope that you and your partner have produced sufficient numbers that have a chance of finding each other. When plants went dry-shod and conquered the land 400 million years ago (p. 41), they preserved much the same strategy. Spores were light enough so they could travel easily on the breeze and the earliest plants cast them to the winds in their millions to increase the odds of successful reproduction. This blunderbuss approach dominated the plant world for the next 350 million years and is still present in many plant groups, such as ferns.

Insects, however, offered plants a more sophisticated option. If a bug can be attracted to visit several plants of the same species successively through the use of an alluring or nutritious flower, the insect can be used to carry pollen directly to a potential mate – manipulative sex by proxy. Instead of investing energy in shedding millions of spores into the air, and relying on chance alone, many flowering plants use animals to ensure their precious pollen makes it to a suitable mate. And yet, the fossil record shows that the flower wasn't an immediate success. The earliest fossil flowers are about 164 million years old, but it wasn't until 100 million years ago that the biological flower-power revolution occurred. However, when it did there was an abrupt increase in the abundance of wonderfully different plants with an amazing array of flowers and seeds. There was also a simultaneous and equally impressive increase in the diversity of flying insects that is unlikely to be coincidental.

The suddenness of this evolutionary radiation, as such increases in diversity are known, was described by Charles Darwin (1809–1882) as an 'abominable mystery' because it did not quite fit his theory of gradual evolution. Since Darwin, massive strides have been made, but there remains much to learn about the early evolution of the flowering plants. Discoveries like this 20-million-year-old, exquisitely preserved flower, *Chaneya*, help palaeontologists understand the rise of complexity and specialization that is so characteristic of the hundreds of thousands of species of flowering plant that we know today.

Acer otopteryx Heer.
Frücht. Oeningen.
vgl. Heer Urwelt X. 336.

Frucht

Spinning samaras
winged maple seed

OFFSPRING HAVE TO MAKE their own way in life. If descendants fail to move away from their parents a species will never expand or be able to take advantage of new opportunities. What's more, parents tend not to want to compete against their own offspring, either for their own sake or for that of their offspring, who carry their genes into future generations. The movement of offspring away from the parent is called biological dispersal, and life on Earth has evolved a fantastic variety of ingenious and successful ways of achieving this.

Animals that can move around in their adult form, such as birds, mammals and fish, have little problem in testing out new grounds before making their own way there. Other animals, such as barnacles (p. 188), which as adults are stuck in one place, use a mobile larval stage to achieve the same thing. Indeed, some larvae can travel thousands of kilometres drifting in strong ocean currents to settle in a new region successfully. Like the barnacle, most plants are firmly rooted into one position from their germination to their death, and their descendants must also successfully reach new grounds. The most straightforward way for a plant to disperse is to send out buds, like the runners of a strawberry plant; this approach is clearly limited by how far the suckers can travel.

The successful conquest of the lands was accompanied by the evolution of methods of spreading spores. Spores are light and travel far, and they take comparatively little energy to produce. A plant can easily make millions, increasing the chance of a successful dispersal. The problem with spores is that they don't include any form of food reserve for the next generation. Offspring are abandoned to the winds without provision. As plants diversified on land, the competition for nutrients, light and space intensified and it became advantageous for offspring to be given a head start in life. Hence the evolution of seeds, which, unlike spores, include a packed-lunch for the next generation. Seeds give the sprouting juvenile enough resources to become rooted and reach above the competition for light. The bigger the seed the more energy is bequeathed and the greater chance the plant has of beating the competition.

One of the most elegant strategies to have evolved is the winged seed, like this 20-million-year-old maple seed from Baden, Germany exceptionally preserved in fine lake sediments. Winged seeds, otherwise known as samaras, start to spin as they fall from their parent tree. Vortices form around the 'wing', giving the seed lift to travel far enough to be entirely independent from its parent and with enough resources to establish a fine tree.

Lackadaisical turtle
Stylemys

IN HIS FABLES, THE GREEK storyteller Aesop made great truths from simple observations. And so it is in palaeontology, where fragments of the past help us recount the history of life on Earth. One of Aesop's most famous and best-loved characters is the slow-moving tortoise and the story of how, against all odds, he beat the speedy and over-confident hare.

Turtles are reptiles. Reptiles evolved from amphibians over 300 million years ago. At this time the world was gradually becoming a drier place, and consequently the massive swamps of the Carboniferous period began to contract. The amphibians, which lay their eggs in water, found this new world much less hospitable. Reptiles, on the other hand, produce watertight shelled eggs that can safely be laid on land without drying out. This gave reptiles a great advantage in conquering the lands. The land-dwelling turtles, including tortoises, have stubby legs and walk slowly. Marine turtles have flippers to allow them to efficiently 'fly' under water, and although more dexterous than their land-based cousins they are far from being the most adept swimmers in the seas. This more lackadaisical approach to life could have been a downfall if turtles had not evolved a hard shell around their body for protection. And what a successful construction it has been. The shells of turtles have remained remarkably similar for over 200 million years, through which time they have endured, and often outlived, the evolution of a steady stream of predators.

Although the shell is anatomically fairly complex, it's not too difficult to envisage how it evolved through extensions and ultimately complete fusion, of the ribs and backbone. However, until recently all complete fossils of turtles were found to have fully developed shells, like this beautiful specimen of the 30-million-year-old tortoise, *Stylemys nebrascensis*. Evidence for the origin of the turtle shell was unknown until palaeontologists recently found the oldest known turtle-like animal, *Odontochelys*, in 220-million-year-old marine sediments in China. *Odontochelys* has a fully developed plastron, or belly shell, but its back had only bony protuberances that had yet to develop into a fully developed carapace. This half-carapace would nonetheless have conferred great protection from predation.

Punctuated equilibrium
Metrarabdotos

AS THE FOSSIL RECORD IS PIECED together, the shape of the tree, or more accurately the bush, of life emerges with ever-greater accuracy. It's clear that the bush is made up of very old branches and sometimes lopped trunks that represent extinctions. But how do new branches spring from the old wood of an established species? Since Darwin (1809–1882), evolutionary biologists commonly thought that over time natural selection meant species changed little by little, generation by generation, as tiny mutations created equally tiny advantages that accumulated in future generations. But in the fossil record most species appear with few intermediaries and evidence of such gradual evolution is scant (but see p. 84). This puzzle confounded palaeontologists and evolutionary biologists for over a century.

Then, in 1972 palaeontologists Niles Eldredge (1943–) and Stephen Jay Gould (1941–2002) drew up a new shape to the bush of life. Instead of seeing the birth of new species as a gradual slow process, they proposed that species remain in a state of evolutionary stasis, unchanging for perhaps millions of years, before a rapid burst of evolution leads to the formation of distinct new species. Evolution in fits and starts. Their theory, called 'punctuated equilibrium', sent ripples through the world of palaeontology and today remains hotly debated.

A pivotal and rigorous assessment of the punctuated equilibrium theory arose from a collection of minuscule fossils quietly resting in a few cabinets in the depths of the Paleobiology department at the National Museum of Natural History in Washington DC. Sliding out the drawers one finds hundreds of small acrylic boxes filled with thousands of minute fossils called bryozoans, which at first glance look like nothing more than tiny broken branches. When magnified under a scanning electron microscope, however, the magnificent beauty of these little aquatic animals (for that is what they are) emerges. Bryozoans, of which there are nearly 6,000 known species, are a boon to the palaeontologist wishing to study evolution because of their intricate forms and great abundance in the fossil record. In the 1980s, paleontologists Alan Cheetham (1928–) and Jeremy Jackson (1942–) embarked on a mission to subject the punctuated equilibrium theory to its most rigorous scrutiny, employing the bryozoans as their test subjects, fully expecting to disprove the theory. Against all their expectations, the minuscule fossils presented compelling evidence that species indeed remained unchanged for millions of years, only to experience rapid evolution leading to the emergence of entirely new species, without any discernible intermediate forms. However, when these same fossils were recently reevaluated using cutting-edge methodologies, the supportive evidence failed to materialize. As the ongoing debate regarding the mechanisms of evolution persists, these unassuming bryozoan fossils, along with their microscopic counterparts, continue to play a central role in deepening our understanding of the astonishing intricacies of the bush of life.

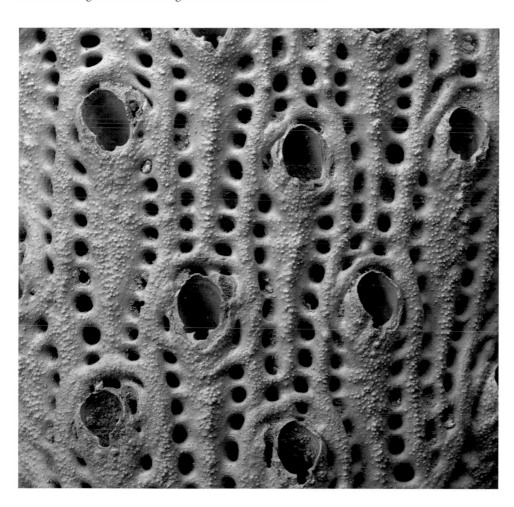

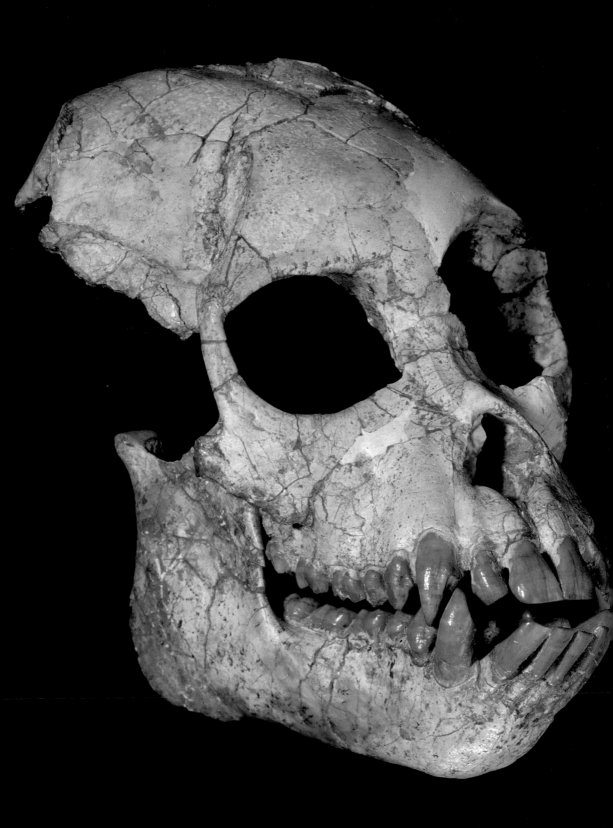

Before Consul
Proconsul

THE SEARCH FOR A COMMON ancestor between humans and their closest living relatives the African great apes (gorilla and chimpanzees) has exercised scientists for more than 100 years. Research has been predictably focused on fossils from Africa, among which is *Proconsul*. This genus was created in 1933 by Arthur Tindell Hopwood (1897–1969), a specialist on fossil mammals at the Natural History Museum, London. Hopwood derived the name from that of a well-known London Zoo chimpanzee of the time called Consul. His description was based on material he had collected personally at Koru in Kenya during an expedition two years previously. However, a gold prospector had actually found the first examples of *Proconsul* there 24 years earlier.

Compared with modern chimpanzees, *Proconsul* was smaller, about 50 centimetres (20 inches) long, and much more monkey-like. The canine teeth were also smaller, and the animal had a long, flexible and curved back. It probably walked along the tops of tree branches rather than swinging from branch to branch. However, like modern apes, *Proconsul* lacked a tail. Its teeth suggest that *Proconsul* fed mainly on soft fruits. The first Kenyan fossils of *Proconsul* were found in rock of Early Miocene age, about 19–20 million years old. Four different species of *Proconsul* have now been recognized, all occurring in Kenya and one in Uganda too. The famous husband and wife team of Louis Leakey (1903–1972) and Mary Leakey (1913–1996) made some of the best finds.

Proconsul was at one time interpreted as a direct ancestor of the African great apes, but this now seems unlikely, not least because the ability to recognize ancestors among fossils with any degree of confidence has been brought into question. Nonetheless, debate continues as to whether *Proconsul* and various other fossil apes of similar age from East Africa are closer to the branch of the evolutionary tree leading to humans or to that including the gorilla and chimpanzees. These fossils are just too old to preserve sufficient DNA that might have provided a definitive answer.

Terrifying twitchers
the terror birds

EVOLVING IN THE JURASSIC PERIOD, 150 million years ago from carnivorous dinosaurs, birds are a magnificent example of an evolutionary triumph (p.103). When one considers the diversity of shapes and colours and ways of life that birds have attained since their reptilian ancestry – from nocturnal hunters to vegetarian divers – it's easy to see why birds generate so much passion in people around the world. But would that same passion be engendered by a massive, flightless bird with a beak capable of killing in a single jab? Luckily for dedicated 'twitchers' more accustomed to observing dazzling plumage and enchanting song, the so-called 'terror birds' are long extinct.

Nearly 20 species of this break-away group of flightless birds have been unearthed, often represented only by a few scattered bones. But now and again, a dedicated palaeontologist discovers a complete skull, like this one of *Patagornis marshi* from Argentina, and it's these skulls that demonstrate the raw power the terror birds must have employed. Computer modelling of skull architecture has revealed that repeated jabbing and pulling back using their powerful neck muscles must have been their preferred method of attack. The massive hooked beak, perfect for fast and lethal stabs, would have been too susceptible to breakage if the struggling prey was shaken from side to side. It is therefore likely that terror birds ruled as top predators by sharp, screamingly fast pecks while their huge three-toed claws held the prey down.

Terror birds emerged in the then island continent of South America shortly after the disappearance of dinosaurs 66 million years ago, and were able to step neatly into the role of top predator left vacant by the likes of *Carnotaurus*, the fast horned dinosaur exclusive to South America. The continent was not only richly populated with mid-sized mammals such as sloths, armadillos and marsupials, which must have made up a good proportion of the birds' diet, but conveniently lacked the array of large fierce mammals well established in North America and Eurasia, such as lions and tigers, that would have stood a chance of fighting back. Indeed, the reason large carnivorous birds only evolved in South America may have been due to the vacant niche that was filled by large fierce mammals elsewhere.

When the Isthmus of Panama rose up 3 million years ago, bridging North and South America, it opened the door to a two-way mass migration known as the 'Great American Biotic Interchange'. Whereas North American migrants such as horses, deer, camels, mastodons and foxes successfully invaded South America, those moving north did not fare so well. The terror birds, examples of which have been found in Texas and Florida, were the only true carnivores to do so. But life north of the border, with its more efficient prey and stiffer competition from other predators, did not suit them, and their sojourn in North America lasted only a million years or so before they were driven to extinction, depriving us of a creature that for all its fearsome size and behaviours would surely have been on any twitcher's must-see list.

In and out of isolation
Thylacosmilus

EACH SPECIES HAS ITS OWN unique way of life, called its niche. Amongst animals on land there are grazers of grasslands and forests, tree-climbing frugivores, predators and parasites. Within these niches there are many further subdivisions, each giving space to the glorious diversity of life.

George Gaylord Simpson (1902–1984), an influential twentieth century palaeontologist, had a knack for using fossils to reveal the niches of long-extinct animals. Once he accepted that continents moved through plate tectonics, a truth he initially resisted, he went on to describe how the animals of the South American continent had evolved separately for most of the last 65 million years. This was because South America was an island that had separated from Australia, Africa and Antarctica but did not join North America until the closure of the Isthmus of Panama just 3 million years ago. During this time of 'splendid isolation' a wonderful array of very unusual animals evolved into very familiar niches. There were giant ground-dwelling sloths that could strip forests bare and strange-looking litopterns that matched the grazing role filled by horses and zebras, even evolving the same one-toed stance. One-tonne rodents had teeth and presumably diets like elephants. And the aptly named terror birds (p. 178) filled the niche of top predator.

The sabre-toothed cat, a ferocious 400-kilogram (882-pound) feline that hunted bison, mammoths and perhaps even humans, employed lethal strikes to the jugular with its two long, curved, maxillary canines. As the sabre-toothed cat evolved from placental mammals in North America a very similar looking animal evolved in South America, not from placental but from marsupial mammals. *Thylacosmilus* raised its young in a pouch outside its body just like kangaroos, but that's where the cuddly similarity ends. As with the sabre-toothed cat, its canines had only one purpose. The role of hunting with sabre teeth had evolved entirely independently on both continents – a phenomenon known as convergent evolution (p. 90).

These fossils beg the question of why the clearly successful niche of predating with sabre teeth is unfilled today. Many scientists suspect that the marsupial *Thylacosmilus* went extinct because the North American sabre-toothed cat, with its greater metabolic efficiency, out-competed *Thylacosmilus* after the Isthmus of Panama joined North and South America. The sabre-toothed cat survived but it too went extinct a mere 10,000 years ago, coinciding with the arrival of humans and the sudden extinction of many of its prey like giant beavers, humpless camels, mastodons and mammoths. One might say that human ingenuity filled the niche of the sabre-tooth.

Fossil frogs
Discoglossus

THERE'S NOTHING QUITE LIKE A FROG – except for a toad, which is basically a frog with warty skin adapted to living in a dry environment. No other animal has ever evolved the same shape as a frog. These tailless amphibians, with their bulging eyes, sloping back and long hind legs designed for hopping, are unique. Strip away the flesh and frogs also reveal many distinctive features in their skeleton. The skull does not have a solid roof, there are typically only six or eight vertebrae, the ribs are very short, and some bones of the limbs are fused or made entirely out of cartilage. At least 5,000 species of frog exist at the present day, far too many of them threatened with extinction through habitat destruction and chytrid fungal infections.

Fossil frogs are rather rare. While most consist of isolated bones, several exceptional deposits preserve complete frogs, such as the superb specimen shown here of *Discoglossus* from the Miocene of Germany. The body is compressed in a mode of preservation referred to by palaeontologists as 'road kill' because of its resemblance to animals that have been squashed under the wheels of motor vehicles. Still alive today, species of *Discoglossus* are known as painted frogs. They resemble common fogs but have pupils shaped like inverted teardrops. Along with the midwife toads, *Discoglossus* belongs to a primitive family of frogs with a fossil record that stretches back to the Jurassic period – so frogs were around at the time of the dinosaurs.

Discoglossids are not the oldest fossil frogs. That honour belongs to *Triadobatrachus*, known from just one Early Triassic specimen collected in Madagascar, which is roughly 240 million years old. Whereas Jurassic discoglossids show most or all of the features seen in present-day frogs, *Triadobatrachus* was rather different. The skull shows it was a frog but the body had a tail and, unlike modern frogs, it may not have been able to jump effectively. Because *Triadobatrachus* shows some but not all of the features of modern frogs it is said to belong to the stem-group of frogs. In the past, it would have been described as an ancestral frog. An even older fossil was discovered recently in the Permian rocks of Texas. *Gerobatrachus* is more primitive than *Triadobatrachus* and may belong to the stem-group of living amphibians, that is the lissamphibians, which include not only frogs but also salamanders. Fossils such as *Triadobatrachus* and *Gerobatrachus* provide invaluable insights into how and when the unique features of frogs evolved in the distant geological past.

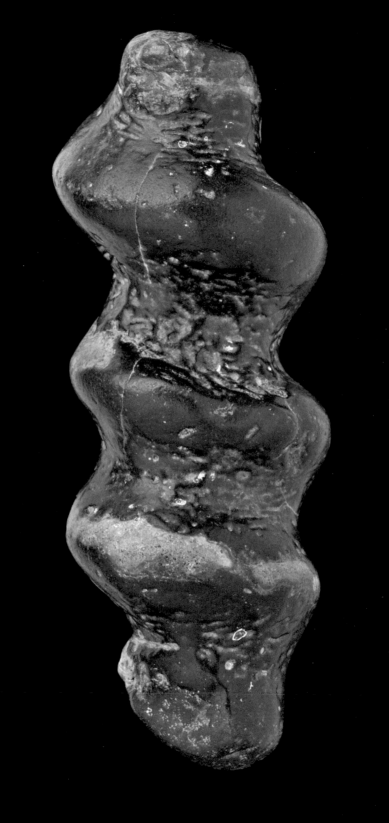

Spiral bezoar
shark coprolite

A BEZOAR IS A HARD CONCRETION or calculus that forms in the intestines of vertebrates. Bezoars grow *in situ* by a variety of different processes, including incremental calcification, like a pearl, or even by the over-consumption of hair. In the seventeenth century, bezoar 'stones' were seen to possess magical powers that most famously included a purported antidote to poison, and they were highly valued and traded widely. In the early nineteenth century, during her excursions in Lyme Regis, the sharp-eyed and sharp-witted fossil collector Mary Anning (1799–1847) (p. 90) found what she called bezoars in the abdomens of some of the fossil ichthyosaurs she had unearthed, but she duly noted that when they themselves were broken open they often contained small fossilized bones and scales of fish.

What Mary Anning had found were the fossilized faeces, or coprolites, of the ichthyosaurs, and the fish remains within them were the remnants of one of the extinct reptile's last meals. To this day, coprolites represent one of the most illuminating types of fossils that can be found, giving us unique insights into the diet and health of ancient animals. Evidence of grass phytoliths in dinosaur coprolites has helped reveal the much earlier than expected origins of the grass family (p. 132). The traces of certain proteins in the archaeological coprolites of ancient humans reveal the telltale signal of cannibalism. And like a stool sample examined by a gastroenterologist, a fossilized coprolite can reveal the parasitic fauna that the unfortunate host was carrying.

Despite their wealth of information, most palaeontologists aren't as lucky as Mary Anning to find the coprolites still embedded within the animal that produced them. It is normally fairly difficult to be sure which animal produced a coprolite, and therefore the usefulness of the information preserved within them can be limited. In some cases, however, there are clever clues that unmask the maker. This 13-centimetre (5-inch) long example of a coprolite was discovered in ancient sea sediments in Suffolk, England and its shape gives it away as having come from a shark. It shares the same distinctive sharply tapering top and bottom like modern shark poo, and only sharks possess a spiral flange in their intestines that leaves the excrement with such a well-defined and narrow spiral groove.

Desert sea
Cyprideis

CALLED THE MARE NOSTRUM, or 'Our Sea' by the Romans, the nearly land-locked Mediterranean Sea has been the centre of trade and wars between empires from the Phoenicians to the Ottomans. Even before historical times, however, the Mediterranean endured ebbs and flows of biblical proportions. Over five million years ago the Mediterranean Sea almost dried up. One might find this hard to believe. The Mediterranean is far from shallow, often steeply falling to over 5 kilometres (3 miles) deep, and its crystal azure waters today give no clue as to its parched history.

So it was when scientists on board the specially designed and innovative deep-sea drilling ship *Glomar Challenger* set to work in 1970 to document the types of rocks that lie on the seabed. The ability to drill into several kilometres of rock ,while floating kilometres above without anchorage, represented a major technical leap forward that would revolutionize our understanding of the history of life on Earth. As Kenneth Hsu (1929–), one of the chief scientists, and the team began pulling up a seemingly incomprehensible jumble of unexpected rock types, an intense struggle ensued in a quest to interpret them. The result was exemplary science leading to a revolutionary understanding.

The first clues to the Mediterranean's arid history came from Drill Hole 124 located between Majorca and Sardinia in waters 2 kilometres (1¼ miles) deep. After drilling through layers of ordinary marine oozes that are deposited in all deep oceans, the drill bit hit a suite of hard rocks that slowed drilling down from metres a minute to less than a metre an hour. When the 'rocks' were hauled to the surface the team were astonished to find both fossilized stromatolites (p. 10) that today grow mostly in shallow water around salty lagoons, and anhydrites – a type of rock that forms only in arid deserts. Shallow lagoons and arid desert at the bottom of the sea. How could this be? The final piece of evidence was the discovery of an unassuming fossil ostracod called *Cyprideis*. This microscopic, shelled crustacean (depicted here), is never found in deep oceans but only lives in shallow salty pools. The only conclusion that could be drawn was that millions of years ago the Mediterranean must have all but dried out.

Calculations reveal that the amount of rainfall entering the Mediterranean through rivers is considerably less than the amount of water that is evaporated from the sea's surface each year. It is only Atlantic Ocean water flowing through the Straits of Gibraltar that keeps the

Mediterranean constantly topped up. If the strait closed, most of the Mediterranean would evaporate within a thousand years, resulting in devilishly hot plains thousands of metres below sea level connecting North Africa with Europe and interspersed with briny lakes and steep mountains.

This is exactly what happened in the past, not just once but many times over a half-a-million-year period. The team aboard the *Glomar Challenger* recovered several cycles of the ooze–stromatolite–anhydrite sequence in Hole 124. Interestingly, the final sequence ended abruptly 5.33 million years ago, suggesting that there may have been a time when, unobserved by Noah or any other human, Atlantic seawater spectacularly cascaded over the Straits of Gibraltar, filling a huge desert thousands of metres below and shaping the crucible of many great empires.

Conserving energy
barnacles

A SHELL CONTAINING TINY holes that he collected from a Chilean beach in January 1835 provided the young Charles Darwin (1809–1882) with a puzzle. Inside the holes were minute animals resembling barnacles, but less complex than the familiar barnacles of the seashore. The passion Darwin was to develop for barnacles in the years following his return from the expedition on HMS *Beagle* is often blamed for the long delay to the completion of his seminal book *On the Origin of Species*. However, Darwin needed to establish his credentials among fellow scientists in order for his theory of evolution to receive serious consideration. At the time, the best way of doing so was to become an expert on the taxonomy – recognizing and classifying species – of a particular group of animals. Most of the other major groups had already been 'bagged', leaving the barnacles for Darwin. By mastering the anatomy of these animals he was also able to understand how the many different species of barnacle were finely adapted to their habitats.

Barnacles are crustaceans related to crabs and lobsters; crustaceans in turn are arthropods and therefore close cousins of other groups with jointed appendages such as insects, spiders and scorpions. There are three basic varieties of barnacle: boring, goose-necked and acorn. Boring barnacles, like those collected by Darwin in Chile, excavate pip-shaped cavities in shells and limestone. Although the delicate body of these animals does not fossilize, the distinctive boring holes can be preserved. The oldest examples of barnacle borings come from the Devonian, showing that barnacles had evolved this lifestyle at least 400 million years ago. Goose-necked barnacles may be an even more ancient group; a possible example has been identified from the Cambrian. Most fossil examples are found as isolated wedge-shaped plates of calcite left after the decay of the stalk and other soft tissues. Finally, acorn barnacles have a volcano-like shell and are enormously abundant today. Yet they are a geologically young group, first appearing close to the end of the Cretaceous period and remaining quite rare until about 20 million years ago.

Barnacles have been likened to shrimps lying on their backs and kicking their legs in the water to collect plankton. However, the larval stages of barnacles are relatively conventional, resembling those of other crustaceans in being free-swimming members of the plankton. Just as the free-swimming barnacle larva at some point settles and cements onto a hard surface to commence its fixed adult life, so in the evolution of barnacles the same transition from a free-swimming ancestor to a fixed descendant occurred. Barnacle ontogeny (development) broadly recapitulates barnacle phylogeny (evolutionary history), a common theme in evolution. From a human perspective, sacrificing freedom to move around for a life tethered to one spot would seem imprudent. But the evolutionary race is not always won by the swift. Instead of expending energy swimming around in pursuit of food, barnacles have evolved the more sedate strategy of allowing ocean currents to bring the food to them.

Left-handed snails
Neptunea angulata

THE HINDU GOD VISHNU holds in one of his four hands a sacred conch shell, known by the scientific name *Turbinella pyrum*. But whereas almost all specimens of this Indian Ocean gastropod are right-handed – or dextral – the one in Vishnu's hand is a left-handed variant, called a 'Dakshinavarti Shankh' by Hindus. The rare sinistral shells are greatly valued, as they are reputed to occur in less than 0.001% of individuals. However, some species of gastropods are routinely sinistral. An example of a sinistral gastropod from the fossil record is *Neptunea angulata*, a species of whelk that is a characteristic fossil in an East Anglian shell gravel called the Red Crag, which is about 2.5–3.3 million years old.

When viewed with the apex of the shell at the top, sinistral gastropods such as *Neptunea angulata* have the aperture on the left, whereas in dextral gastropods the aperture is on the right (p. 160). The two shell types are mirror images. Gastropod shells grow as helical spirals, anticlockwise in sinistral gastropods but clockwise in dextral gastropods.

With no obvious advantage in being sinistral as opposed to dextral, it might be expected that the frequency of species showing each coiling direction would be roughly half and half. In fact, more than 90% of modern gastropod species are dextral, with the few sinistral species mostly living in freshwater environments. Evidence from fossils shows quite clearly that dextral species have predominated over sinistral species ever since gastropods first appeared in the Cambrian. Why should this be? This is a question that has long taxed the minds of biologists and has yet to be convincingly answered.

According to an estimate made by evolutionary biologist Geerat Vermeij (1946–), sinistral species of marine gastropod have evolved independently from dextral ancestors 19 times during the last 70 million years. Vermeij believes that the transition from dextral to sinistral is more likely to occur in habitats where few predators are present. This is because predators normally remove the first sinistral mutants, which tend to show abnormalities and weaknesses in addition to their opposite coiling.

In some gastropods sinistral mutants have difficulty mating with dextral individuals that have reproductive organs on the opposite side of the body. But when more than one sinistral mutant is present, the sinistral individuals may mate only with one another, instantly creating the kind of reproductive isolation necessary for the formation of a new biological species.

Bunyip at the billabong
giant wombats

EVERY CULTURE HAS ITS mythical creature ready to devour children should they roam too far or fail to come home in time. Invariably they are bloodcurdling animals that lurk in places where too much fun could be had. For the Aboriginal people of Australia the Bunyip fills this role. Like most mythological beasts the Bunyip comes in a manner of guises, and has been described as a monster with flippers, feathers and a dog-like face with supernatural powers. Most accounts concur that the Bunyip is a hideous creature the size of a cow that frequents watering holes and rivers.

So, when early surveyor of southeastern Australia, Sir Thomas Livingstone Mitchell (1792–1855), discovered a cache of large fossil bones in a cave in Wellington in the early 1830s the evidence was at hand that the Bunyip was a real creature that inspired the myth. Esteemed British palaeontologist Richard Owen (1804–1892) was the first to describe the bones as belonging to a giant marsupial and named it *Diprotodon*, meaning 'two forward teeth' for its remarkable dentition. Since then, many bones of this giant rhinoceros-sized wombat have been discovered, some with juveniles associated with a female adult in the position where a pouch would have existed, just like the wombats and kangaroos – attesting to it being a true marsupial.

Diprotodonts were vegetarian giants that roamed the entire continent. Fossil hoards suggest that they formed strong family communities with clear gender segregation, like modern elephants, where the females and young formed caring groups and males roamed and fought for the right to mate with the females. The oldest fossils appeared around 1.5 million years ago, and they abruptly disappeared around 50,000 years ago. The cause of their disappearance is hotly debated. Clusters of *Diprotodon* bones dominated by young and old animals suggest mass deaths caused by drought because the weakest are more likely to die. As Australia became drier over the last few million years, the extinction of the diprotodons may have become more and more likely. Others point out that their disappearance was abrupt and not gradual and so cannot have been due to climate change, but in fact neatly coincided with the arrival of humans to the world's smallest continent.

Indeed, upon arrival, humans immediately began to alter the vegetative landscape of Australia through the use of fire, which may have indirectly led to the extinction of *Diprotodon*, as their habitat became scarcer. More damning evidence is that some bones have been found with possible butchery marks. There is little doubt that capturing a *Diprotodon* would have been a relatively easy task and a single animal would have provided enough for a communal feast. *Diprotodon* wasn't the only casualty. Giant wallabies and kangaroos also disappeared at this time. The Australian wildlife is a meagre representative of what it once was, and the job of convincing an Internet-savvy child not to linger by the billabong a little harder.

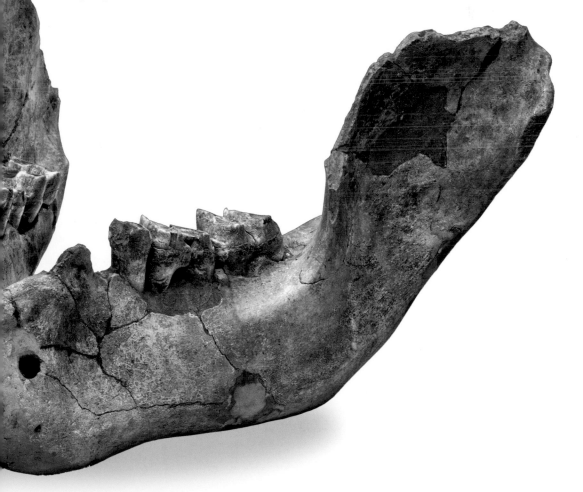

American interchange
glyptodont

REACHING THE SIZE OF A Volkswagen Beetle, and not dissimilar in shape, the shield of the extinct glyptodonts served the same protective purpose as the carapace of a tortoise. Unlike the tortoises, however, glyptodonts were unable to retract their head into their shell and instead relied upon a shielded skull for protection. Some glyptodonts, such as *Hoplophorus*, took defence one-step further with a modified tail tipped with an intimidating club of spikes. Along with their smaller and less heavily armoured cousins, the armadillos, glyptodonts were mammals that evolved in South America – a continent isolated from other continents for 66 million years until the rise of the Isthmus of Panama 3 million years ago paved the way for a massive two-way migration called the 'Great American Biotic Interchange'.

Intriguingly, most North American migrants, including foxes, bears and camels (llamas), not only settled but flourished in South America, whereas the animals from South America were much less successful at both invading the new lands of North America and coping with the northerly invaders. Many suspect that this was because the northern fauna had evolved with a consistent connection to Eurasia and it had therefore already endured, and survived, many different types of competition. The fauna of North America was in essence highly honed while the South American fauna had evolved in isolation for 66 million years, and as such had never before seen the intense competition it would experience from its northerly neighbours when the land bridge formed. The result was widespread extinction among the fauna from South America.

However, the glyptodonts were one of the exceptions to the rule. They initially survived in abundance, perhaps because of their colossal investment in protection. Yet, what might have safeguarded the glyptodonts for millions of years may have ultimately led to their demise. Glyptodonts went extinct around 10,000 years ago, coinciding with the arrival and spread of the earliest humans into the Americas, and there is evidence that these early Americans used the enormous carapaces of the glyptodonts as shelters or dwellings. We do not know for sure if they were killed specifically for this purpose, but once again nimbleness and ingenuity gave man the advantage with lasting consequences.

Missing links
Australopithecus

THE QUEST FOR EVOLUTIONARY missing links has been pursued more vigorously in palaeoanthropology than in any other branch of palaeontology. Until quite recently, the expectation was that a single line would lead from an ape-like ancestor to *Homo sapiens* – almost any newly discovered fossil hominin (the name given to fossil species thought to be closer relatives of humans than to the living great apes) could be placed somewhere along this lineage and legitimately regarded as a missing link. The idea of a single lineage leading to modern humans has become ingrained in popular culture by the famous and much parodied 'march of progress' illustration drawn by Rudolf Zallinger (1919–1995) for a 1965 book on early humans. This shows the profiles of a line of 15 hominins, beginning with a crouching, long-armed ape on the left and ending with a tall, upright human on the right. The 'ape-men' between are the missing links.

One of the best-known 'missing links' is *Australopithecus africanus*, first discovered in South Africa in 1924. The publication announcing the discovery of *Australopithecus* by Raymond Dart (1893–1988), an anatomist at the University of the Witwatersrand, was met with a certain degree of scepticism by palaeoanthropologists, many of whom believed it to be the fossil of an ape rather than a close human relative. Dart defended the importance of *Australopithecus*, noting that it vindicated Darwin's claim that Africa would prove to be the cradle of mankind. Although Dart has been proven correct in most respects, the plethora of fossil hominins discovered in recent years have altered our view of human evolution: the 'march of progress' iconography is incorrect and *A. africanus* can be viewed as one of numerous branches on the hominin evolutionary tree, not as a direct ancestor of modern humans.

Numerous species of *Australopithecus* have since been found in different parts of Africa such as Chad, Ethiopia, Kenya and Tanzania. But the defining species for the genus remains Dart's *A. africanus*. Most of the thousands of fossils that have now been collected of this species come from ancient caves near Johannesburg. Along with other animals, individuals of *Australopithecus* were either washed into the caves, fell in through holes in the roof, or were dragged there by carnivores. Dating the fossil bones from these caves can be difficult, although it is believed that most of those of *Australopithecus* are 2.0–4.5 million years old. No stone tools are demonstrably associated with *A. africanus*; unlike younger species placed in the genus *Homo*, *A. africanus* may not have manufactured stone tools.

Australopithecus africanus can be shown to have been bipedal from various elements of the skeleton, such as the pelvis and femur, and the basal position of the hole in the skull where the backbone was attached. Males are estimated to have stood up to 140 centimetres (55 inches) tall, and females about 115 centimetres (45 inches). Weighing an estimated 450 grams (just under a pound), the brain of *A. africanus* was only slightly larger than that of a chimpanzee (395 grams/a little over ¾ pound), and far smaller than an average modern human (1,350 grams/just under 3 pounds). The fingers were curved and ape-like, indicating tree-climbing ability; the face was flatter and more human in appearance than that of a modern ape; and the canine teeth were human-like in their smaller size.

Companions in cinder
the Laetoli footprints

IT'S ONE OF THE MOST EMOTIONALLY engaging stories in palaeontology. Over three and a half million years ago a volcano in Tanzania smothered the savanna in an ash that was ideal for preserving the imprints of creatures that passed over it. Amongst the criss-crossing trails of hares, hyenas and a small extinct horse with foal that are preserved are some unmistakably hominin tracks. They represent not just one but at least three individuals travelling in the same direction, and all walking upright on two legs.

The 27-metre (88 foot) long hominin trails were discovered in 1976, excavated by Mary Leakey (1913–1996) near the village of Laetoli and are known as the Laetoli footprints. There is little doubt that they were made by an early human, as the feet that made them had a big toe that was in line with the foot – a characteristic that separates humans from apes, who have splayed big toes more useful for climbing trees. The heels dug deeper than the toes, implying a fully upright human-style of walking rather than the bent-knee and bent-hip gait so characteristic of non-human apes. Who made the Laetoli prints? The early hominin *Australopithecus afarensis* is the only contemporaneous hominin known to date. Before the discovery of the footprints a debate raged about which came first in human evolution: bipedalism or a large brain? *Australopithecus* had a small brain (p. 197), less than half the size of a modern human, yet here was evidence that they walked upright. Bipedalism must have evolved first, and long before brains expanded and complex tools were fabricated.

As one looks at the tracks, we can try and piece together the sequence of events. The two clearest trails are made by a smaller and a larger individual, perhaps a male and a female, and they are so close that if they had been together they must have had to walk arm in arm. The prints of the smaller individual suggest it was carrying something heavy on one side, perhaps a child. There is a third set of prints that were laid down inside the first larger prints. Could they have been made by an adolescent, stubbornly following the father, and represent evidence of a family escaping an erupting volcano? These suggestions are as emotive as they are speculative. Yet, despite the uncertainties as to how they were made, we can still marvel at the improbability of their preservation and discovery and accept the unassailable significance that bipedalism evolved very early. These trace fossils are not the dry remains of a long extinct animal, but a record of active life with a real connection to us. They show people doing what people do best – travelling, and possibly together, three and a half million years ago. No wonder our imagination runs wild.

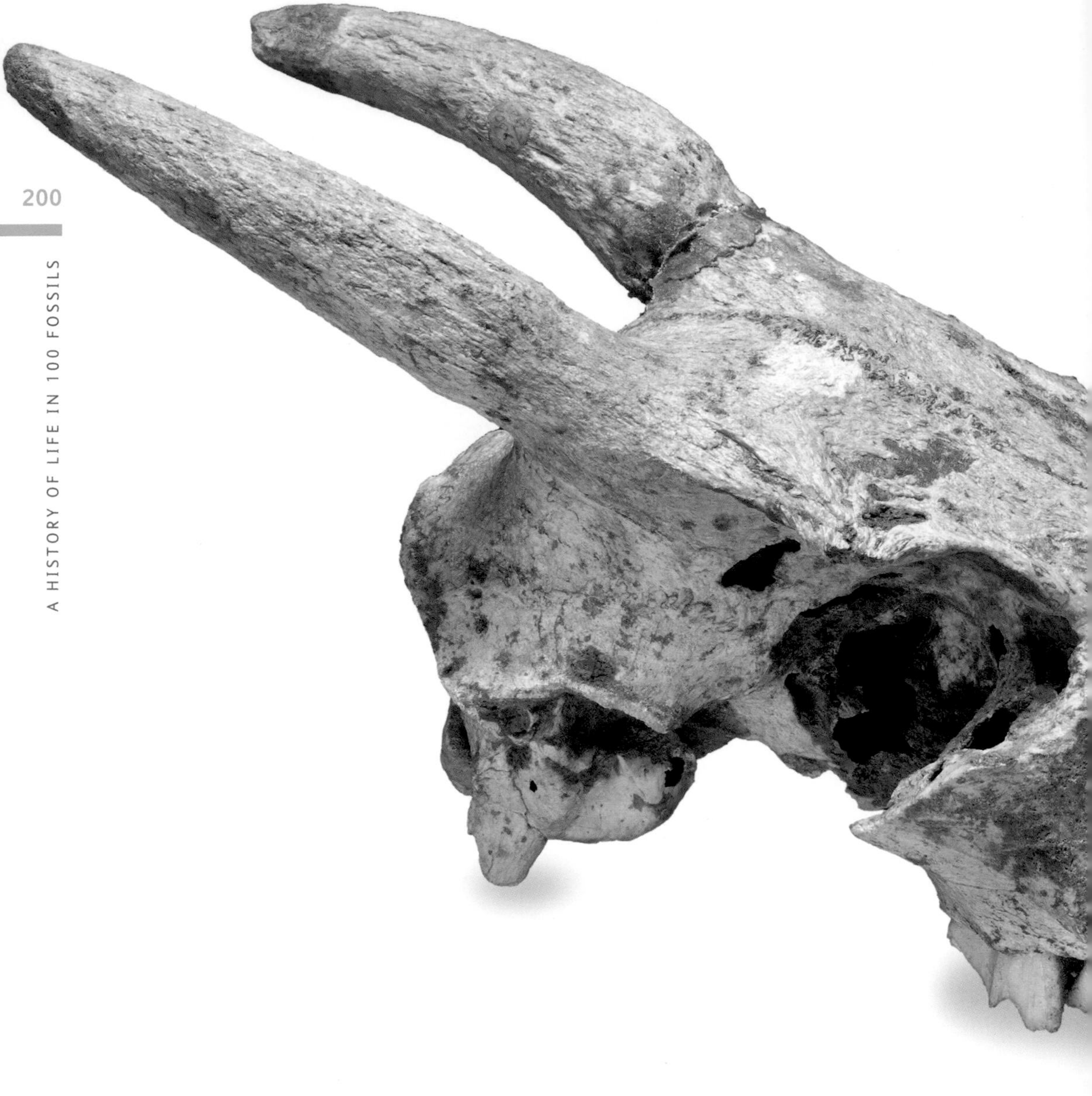

Island dwarfism
the mouse goat of Minorca

NORMALLY IT HELPS TO BE BIG if you are an animal: you are less likely be eaten, more capable of subduing animals you wish to eat, and – like the giraffe – you will be able to reach foliage and fruit on high branches. But large size brings with it one major disadvantage: big animals require a lot of food. So when species of a large size colonize small islands they often exhibit dwarfism. In these settings, natural selection may favour small body size, because food resources can be limited. Also, because predators tend to be fewer on islands, one of the main benefits of being big vanishes.

In addition, the numbers of animals that can live on a small island is limited. This so-called 'carrying capacity' of the environment is lower for large animals: there are far fewer elephants than mice. And small populations are more vulnerable to extinction – a fact well known to conservation biologists. Dwarf species are able to maintain larger population sizes, meaning that they are less likely to become extinct.

A natural stage for the evolution of dwarfism was provided by some Mediterranean islands during the Pleistocene epoch. Fossils of pygmy elephants and deer have been found on Cyprus, Crete, Malta and Sicily. Further to the west, the Balearic islands of Minorca and Majorca were home to a pygmy goat until about 4,500 years ago. Bones of this animal – the so-called 'mouse-goat', *Myotragus balearicus* – were first discovered by the pioneering fossil hunter Dorothea Bate (1878–1951). Adults stood about 50 centimetres (20 inches) tall at the shoulder and, for their size, had a proportionally small brain and short legs – it seems there was little need to be especially intelligent or fleet of foot in a natural environment containing few predators. Fossil faeces (coprolites) found in caves occupied by mouse-goats indicate that the diet of the animals included the toxic Balearic box, *Buxus balearica*. But eating this shrub was not the reason for their extinction; rather, the culprits were probably humans, perhaps in combination with climate change. Before they had been overhunted, however, unsuccessful attempts had apparently been made to domesticate mouse-goats. This is indicated by the occurrence of fossil mouse-goats with trimmed horns showing regrowth – such trimming could only have been accomplished by humans, and it clearly happened during the lifetimes of the animals.

Directional evolution
Megaloceros

NO MODERN DEER BOASTS antlers more magnificent than those of the extinct giant deer *Megaloceros giganteus*, or 'Irish elk' as it is often misnamed. The antlers of this relative of fallow deer had a span of up to 3.5 metres (11 feet 6 inches). Many mounted examples are to be seen not only in museums but also adorning the walls of stately homes, together with the far punier antlers of modern species of deer. The living deer probably stood about 2 metres (6 feet 6 inches) high at the shoulders.

This species of giant deer lived between about 450,000 and 7,700 years ago and could be found from Ireland in the west to central Siberia in the east. The reason for its extinction is nowadays usually blamed on climatic change, hunting by humans or a combination of the two. However, these theories have not always prevailed. At one time, the extinction of *M. giganteus* was considered to be the natural end product of orthogenesis ('directional evolution'). Briefly in vogue during the late nineteenth and early twentieth centuries, orthogenesis was a theory that explained evolutionary change in terms of internal 'compulsions' driving evolution along predefined pathways. Once particular features had evolved, they were apt to become more extreme with time, eventually becoming so disadvantageous that the species possessing them was driven to extinction. The huge antlers of *M. giganteus* seemed to fit this idea particularly well, appearing to be so hefty that the deer would have great difficulty holding its head up, the antlers becoming mired in the boggy ground over which the animal roamed or entangled in the branches of trees. While orthogenesis has been thoroughly discredited as an evolutionary theory, the question still remains as to why the antlers were so large in *M. giganteus*. What function might they have served? Only the males (stags) of the species grew antlers, leading to the conclusion that the antlers are a product of sexual selection. Dominant males (with larger antlers) are more likely to mate with females and pass their genes to the next generation. In fact, the antlers in *M. giganteus* are not particularly over-sized when the large size of the body as a whole is taken into account, and it seems likely that they were used for display purposes and ritualized combat between males, only occasionally leading to serious fighting and injury. The huge antlers of male individuals of *M. giganteus* may not have been as maladaptive as formerly assumed. Some scientists now believe that problems faced by females of the species led to the extinction of *M. giganteus*. Climatic change at the end

of the Pleistocene is thought to have shortened the growing season of the plants consumed by *M. giganteus*, lessening the period when they could be digested and placing critical nutritional stress on pregnant and lactating females.

Mammoth tales
steppe mammoth

AS GARGANTUAN ICONS of the Ice Age, mammoths excite almost as much interest as dinosaurs. But whereas dinosaurs had long been extinct before humans appeared on Earth, mammoths and humans did coexist. Indeed, there is some evidence that humans hunted the woolly mammoth, *Mammuthus primigenius*, to extinction, although the loss of favourable habitat through climatic warming probably also played a role.

The woolly mammoth is one of about 10 different species of mammoth to have lived between about 5 million and 4,000 years ago in Asia, Europe and North America. One of its predecessors, the steppe mammoth, *M. trogontherii*, was among the largest of all mammoths, standing up to 4 metres (13 feet 1½ inches) high at the shoulder. This superb example of the skull of a steppe mammoth, with tusks almost 3 metres (9 feet 10 inches) long, was unearthed in 1863. It was found in a brick-pit at Ilford, then a village but now one of the outerlying boroughs of northeast London, and excavated by English amateur naturalist Sir Antonio Brady (1811–1881). The Ilford mammoth lived during one of the warmer, interglacial periods, about 200,000 years ago. Compared with the woolly mammoth, it was probably less hairy and had teeth with fewer enamel ridges. These were suited to eating the leaves of trees and shrubs as well as the tough grass that came to be relied upon by the woolly mammoth in colder climates.

Mammoths are close relatives of today's elephants, with which they are classified in the family Elephantidae. In fact, it is likely that mammoths evolved from the same ancestor as the modern Asian elephant, *Elephas,* after its evolutionary split from the African elephant, *Loxodonta*. The value of DNA sequence data in recognizing different species and how they are interrelated is well illustrated by elephants and mammoths. Until recently, only one species of African elephant was distinguished, but analysis of DNA suggests that there are two, the savanna elephant, *L. africana*, and the forest elephant, *L. cyclotis*. The relatively young age of mammoths in geological terms, as well as their frequent preservation as frozen carcasses in permafrost, has allowed researchers to recover DNA from these fossils. For example, a recent study of the DNA from two species of mammoth living in North America, the woolly mammoth and the Columbian mammoth, *M. columbi*, has suggested that they interbred.

Lancewood legacy
giant moa

ON THE EVENING OF Tuesday 12 November 1839, a broken bone 15 centimetres (6 inches) long and with a diameter almost half as great was displayed at a meeting of the Zoological Society of London. The exhibitor was Richard Owen (1804–1892), inventor of the term 'dinosaur', later to become first Superintendent of the British Museum (Natural History), and a renowned anatomist even when in his mid-30s. Owen had received the bone, a femur shaft, from a Mr Rule in New Zealand. An accompanying letter stated that the locals believed it to have come from an extinct, eagle-like bird, which they referred to, bizarrely in today's parlance, as a 'movie'. The state of preservation of the bone indicated to Owen that it was of no great geological antiquity. Although Owen agreed that the bone was indeed from a bird, his knowledge of avian anatomy allowed him to identify the source as a giant ostrich-like bird. This was the first of hundreds of thousands of moa bones known to science.

Moa were enormous flightless birds native to the islands of New Zealand, until their final disappearance about 600 years ago. Hunted to extinction by humans, the only previous predator of moa had been the equally huge Haast's eagle. This too was driven to extinction once its major source of food had gone. Fossil bones and eggs point to the existence of not one but as many as 9 different species of moa in the fairly recent geological past. The largest grew to 3.5 metres (11 feet 6 inches) in height, and at least some were characterized by extreme sexual dimorphism – females could be 50% taller than males of the same species. Examples of moa trackways in mud preserve vivid impressions of the oversized, three-toed feet of the giant birds.

These colossal cousins of the kiwi were herbivores, as indicated by the contents of fossilized gizzards preserving pieces of trees and shrubs shredded by their powerful beaks. Remarkably, the native plants of New Zealand today betray the former presence of moa on the islands. Several plant species show features interpreted as adaptations against being eaten by moa. For example, the leaves of lancewoods are initially stiff, spiny, brown and apparently unappetizing, before becoming green and broader when the plant has grown beyond the reach of the moa that once browsed them. Thus the legacy of the moa persists, hundreds of years after their extinction, in the plants they once ate.

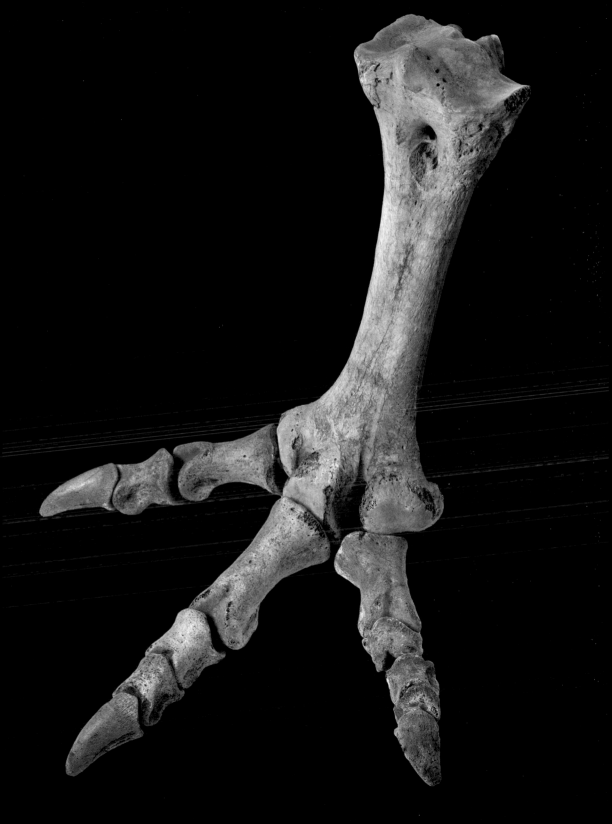

Staghorn coral

Acropora cervicornis

ARISTOTLE IN THE FOURTH century BC described the use of a metal cauldron for use as a diving bell to allow people to breathe while walking on the sea floor. By around 1300, Persian divers were using diving goggles made from polished tortoiseshell. Despite this long history of exploring the ocean floors, it has only been during the last 40 years or so that marine biologists have started to monitor life on the sea floor in any detail. Since then, however, we have witnessed some striking changes. One of the most shocking has been the rapid decline in living corals across the Caribbean.

When scientists began to study Caribbean corals, the fast-growing branching staghorn coral, *Acropora cervicornis*, pictured here, was the most abundant reef-building animal. Today it is rare to find a living example, and the species is decorated with the status of being in critical danger of extinction by the International Union for Conservation of Nature (IUCN). So quick and recent was its decline that scientists still debate the reasons why. Some blame white band disease, which affects only these types of corals, moving slowly up the coral branch degrading the living tissue and killing the coral. Coral bleaching, which is the expulsion of the symbiotic algae that live with the coral has also been accused, although coral bleaching does not necessarily result in the death of the coral animal. Others think the cause of the decline may be less direct. Overfishing, for example, may be important because when we take too many fish out of the water fleshy algae start to grow out of control and smother the corals – this is because the fish that normally keep them in check are no longer there.

Until very recently this transformation of Caribbean reefs was obvious for all to see. The floors of reefs that were once filled with living staghorn coral were littered with the dead skeletons: a striking souvenir of the reef that once existed. But now even those skeletons are being buried by sands and first-time divers on the reef have little to tell them of the glorious past. In some places, however, thick fossil reefs made up entirely of staghorn corals are fortuitously preserved. They stand as an arresting reminder that after dominating Caribbean reefs for over two million years, the staghorn coral is now in danger of disappearing forever.

Sirenian slaughter
Steller's sea cow

IN 1741, CAPTAIN VITUS JONASSEN BERING (1681–1741), an explorer and officer in the Russian Navy, on orders from Peter the Great, led an expedition to map the Alaskan coast. Bering's ship became shipwrecked on what was to become Bering Island and Bering himself died there of scurvy. Half the crew, including German-born naturalist Georg Wilhelm Steller (1709–1746), survived thanks to the discovery of an extremely large sea cow. The meat from these animals sustained the crew, and they were able to build a small ship from the wreckage and return to Russia. During the return voyage Steller devoted himself to documenting the animals and plants they had discovered. The sea cow was of particular interest, not least because without it, he and the remaining crew would have perished. Up to 8 metres (26 feet 3 inches) in length and 10 tonnes (11 short tons) in weight the beast was considerably bigger than any of the manatees and dugongs previously seen.

All sea cows are members of the order Sirenia – marine mammals more closely related to elephants than cows. Described by Steller as having thick black skin 'like unto the bark of an ancient oak', a small head, stubby forelimbs and a whale-like fluke, *Hydrodamalis gigas* was observed to float on the water surface, scraping away at kelp, its principal source of food. Once back in Russia the virtues of the sea cow's meat, milk and hide (useful for repairing boats) were quickly communicated by the returning crew, and sea otter fur-trading expeditions armed with harpoons were instructed to make use of this convenient depot en route. The sea cows were found to be naturally docile and they were easily hunted, providing a wealth of food and profit for the hunters travelling the inhospitable Bering Sea. The sea cows were also famously monogamous. A wretched story told by Steller himself recounts how when one cow was hunted the bull not only tried to save his partner by butting the boat and trying to remove the harpoons, but remained close to the corpse for a day and even the entrails for another two days. Subsequent hunters were unmoved by these touching displays and within 27 years of being described by Steller, *H. gigas* was driven to extinction.

The fossil record reveals a deeper and perhaps darker history. Fossils of *H. gigas* have been discovered from Japan to Mexico, demonstrating that the animal once wallowed in the vast kelp beds around the entire north Pacific Rim. The small populations discovered by Steller must have been the last remnants of a once much larger and healthier population, and it is likely that aboriginal hunting had already reduced them to near extinction levels before the final blow that came from the Western hunters. The virtues of *H. gigas* clearly preceded Bering's expedition. Although the living close relatives of *Hydrodamalis* are rarely hunted today, all four of the much smaller manatees and dugongs are still under threat of extinction themselves.

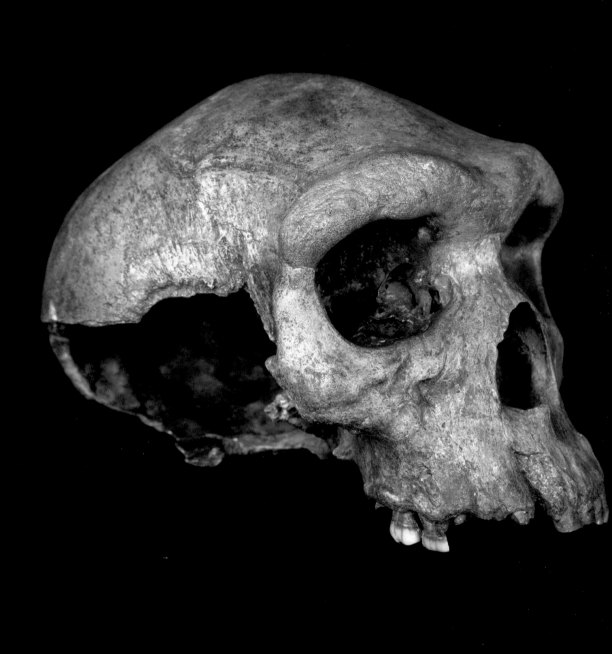

Our last ancestor?

Homo rhodesiensis

THIS MAGNIFICENT HOMININ fossil is the skull of Broken Hill Man, found in 1921. It was the first significant hominin fossil ever to be discovered in Africa and is one of the best preserved of all such fossils. Swiss miner Tom Zwigelaar discovered it in a lead and zinc mine near what is now Kabwe in Zambia. The skull belonged to an adult male and is estimated to be between 275,000 and 325,000 years old. Although undeniably human in general appearance, the skull has a huge face and eyebrow ridges, and a low forehead.

The Broken Hill skull was originally described as *Homo rhodesiensis*, but until recently scientists believed that it belonged to the same species as *H. heidelbergensis*, which was described in 1908 from Mauer in southwestern Germany. This species has also been found in locations such as southern Britain, Greece and Ethiopia. The oldest European and African fossils of *H. heidelbergensis* are at least 600,000 years old, which, with the Broken Hill fossils, suggests that species of this kind existed for more than 300,000 years.

Males of *H. heidelbergensis* may have averaged 1.75 metres (5 feet 9 inches) tall and females 1.57 metres (5 feet 2 inches), close to the stature of modern humans. They were heavily built and clearly capable of surviving in cooler climates than older hominins confined to Africa. The capacity of the skull points to a brain averaging about 1,200 grams (just over 2½ pounds) in weight, roughly 14% lighter than that of an average modern human. A large variety of tools have been found in association with bones of *H. heidelbergensis*. These include various faceted handaxes and possibly also the remains of wooden spears. The Broken Hill site produced scrapers and possible bone tools as well as a remarkable spherical lump of granite. The latter may have been used for pounding food or even been tied to other spherical stones with string and used as a bola in hunting to entangle the legs of prey animals. Regrettably, however, the now exhausted Broken Hill mine was complex, contained at least two caves, and was never properly excavated, leaving open the possibility that the spherical stone had no connection with *H. heidelbergensis*.

Homo heidelbergensis has particular importance in human evolution, as many scientists interpret it as the common ancestor of modern humans, *H. sapiens*, and Neanderthals, *H. neanderthalensis*. If so it currently occupies the last dividing point on the human branch of the great evolutionary tree of life, at least until a new fossil or interpretation comes along.

p.8
Primaevifilum delicatulum
Precambrian (Early Archaen),
  Marble Bar, Western Australia
NHMUK V63164 [6]
Fossil 45 μm across

p.11
Collenia sp.
Precambrian (Early Proterozoic),
  Minnesota, USA
Anton Kearsley Collection
9 cm across

p.12
Tianzhushania sp.
Ediacaran, Guizhou Province, China
Swedish Museum of Natural
  History
c. 700 μm across

p.15
Dickinsonia costata (right) and
Dickinsonia tenuis (left)
Ediacaran, Ediacara Member of
  Rawnsley Quartzite, Ediacara
  Hills, South Australia
South Australian Museum,
  Adelaide P49354
Larger individual 7.5 cm; smaller
  individual 4.5 cm across

p.19
Anomalocaris canadensis
Cambrian, British Columbia,
  Canada
The Geological Survey of
  Canada GSC 75535
11 cm high

p.20
Hallucigenia fortis
Cambrian, Mafang, China
RCCBYU 10248
Fossil 8.2 mm long

p.23
Protocinctus mansillaensis
Cambrian, Purujosa, Spain
NHMUK EE14916
Fossil 1 cm across

p.25
Didymograptus gibbenilus
Lower Ordovician, Meredith,
  Victoria, Australia
NHM H3981–2
7.7 cm across

p.26
Euryeschatia reboulorum
Ordovician, Morocco
NHMUK EE13857
12 cm across

pp.28–29
Orthoceras regulare
Ordovician, locality unknown
NHMUK C4069
17 cm long

p.31
Macrostylocrinus and
platyceratid
Silurian, Middleport, New York, USA
NHMUK EE15429
Crinoid 9 cm across

p.33
Conodonts sp.
Silurian, Gotland, Sweden
NHMUK (no registered
    number)
Field of view 3.7 mm across

p.35
Halysites catenularia
Silurian, Dudley, UK
NHMUK 56534
Field of view 2 cm across

p.37
Gissocrinus goniodactylus
Silurian, Dudley, UK
NHMUK E45528
18 cm across

p.38
Pterygotus anglicus
Devonian, Arbroath, UK
NHMUK I12035
Small individual
55 cm long

p.40
Cooksonia pertoni
Lower Devonian, Herefordshire, UK
NHMUK V58007
Fossil 10 mm high

p.42
Cephalaspis lyelli
Devonian, Forfar, UK
NHMUK P20087
10.5 cm across

pp.44–45
Erbenochile erbeni
Devonian, Morocco
NHMUK It27126
5 cm across

pp.46–47
Comura sp.
Devonian, Morocco
NHMUK It26538
10 cm across

pp.48–49
Mucrospirifer thedfordensis
Middle Devonian, Thedford,
   Ontario, Canada
NHMUK BB58852–9
1.8 cm across

pp.50–51
Cheiracanthus murchisoni
Devonian, Banffshire, UK
NHMUK P4615
Fossil 7.5 cm across

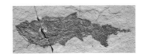

p.52
Pterichthys milleri
Devonian, Scotland, UK
NHMUK 50109
12 cm across

pp.54–55
Eusthenopteron foordi
Devonian, Miguasha, Canada
NHMUK P15956
25 cm across

p.57
Paraconularia quadrisulcata
Lower Carboniferous, Glasgow, UK
NHMUK PG4480
6.7 cm across

p.58
Lepidodendron aculeatum
Upper Carboniferous, locality
   unknown
NHMUK V297
Field of view 12 cm across

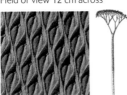

p.61
Glossopteris indica
Permian, Magpur, India
NHMUK V19617
12 cm across

p.62
Eryops megalocephalus
Permian, Texas, USA
NHMUK R6715
30 cm across

p.65
Helicoprion sp.
Lower Permian, Idaho, USA
USNM V22577
26 cm

pp.66–67
Dimetrodon grandis
Early Permian, Texas, USA
USNM V8635
3.1 m

p.68
Cyclacantharia kingorum
Permian, Hess Canyon, Texas, USA
NHM BB11996–9
5.5 cm across

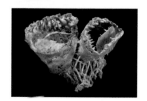

pp.70–71
Deltablastus jonkeri
Permian, Timor, Indonesia
NHMUK E59789–59812
6 cm across

p.75
Tiny Triassic snails
Triassic, Val Brutta, Italy
NHMUK (no registered number)
Thin section, field of view 9 mm
  across

pp.76–77
Massetognathus sp.
Middle Triassic, La Rioja Province,
  Argentina
NHMUK (no registered number)
Skull 18 cm long

p.79
Megazostrodon rudnerae
Upper Triassic, Pakane, Lesotho
NHMUK M26407
3.7 cm across

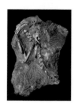

pp.80–81
Clevosaurus sp.
Triassic, Mendips, UK
NHMUK (no registered number)
Skull *c.* 3.5 cm across

p.83
Dapedium sp.
Lower Jurassic, Lyme Regis, UK
NHMUK P6
17.5 cm across

pp.84–85
Gryphaea sp.
Lower Jurassic, Dorset, UK
NHMUK L360
15.5 cm across

p.86
Promicroceras planicosta
Lower Jurassic, Somerset, UK
NHMUK C49680
Field of view 6 cm across

p.89
Cenoceras astacoides
Lower Jurassic, Whitby, UK
NHMUK C35095
13 cm across

pp.90–91
Ichthyosaurus communis
Lower Jurassic, Lyme Regis, UK
NHMUK 36256
75 cm across

pp.92–93
Plagiophthalmosuchus gracilirostris
Lower Jurassic, Whitby, UK
NHMUK 59208
80 cm across

pp.94–95
Muraenosaurus sp.
Middle Jurassic, UK
NHMUK R2421
63 cm across

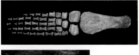

p.96
Neosolenopora jurassica
Middle Jurassic, Chedworth, UK
NHMUK V60738
18 cm across

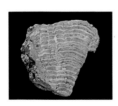

p.99
Leedsichthys problematicus
Middle Jurassic, Peterborough, UK
NHMUK P10000
3 m tall

p.101
Pictonia baylei
Upper Jurassic, Wiltshire, UK
NHMUK C631
11 cm across

p.102
Archaeopteryx lithographica
Upper Jurassic, Solnhofen,
    Germany
NHMUK 37001
60 cm wingspan

p.105
Turanophlebia sp.
Upper Jurassic, Solnhofen,
    Germany
NHMUK 46336
10 cm across

p.106
Mesolimulus sp.
Upper Jurassic, Solnhofen,
    Germany
NHMUK (no registered number)
Fossil 40 cm long

p.109
Rhamphorhynchus muensteri
Upper Jurassic, Solnhofen,
    Germany
Cast at USNM V71
Original at YPM VP.001778
Distance from wingtip to wingtip
    45 cm

p.110
Iguanodon sp.
Lower Cretaceous, Tilgate,
    Sussex, UK
NHMUK 36497
4 cm across

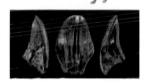

p.113
Female *Microphorites* fly
and nymphal oribatid mite
in silk strands of orb spider
Lower Cretaceous, San Just,
    Teruel Province, Spain
CPT-963 and CPT-964, Fundación
    Conjunto Paleontológico de
    Teruel, Spain
Main piece 8.2 x 4.8 mm

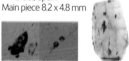

p.114
Araucaria mirabilis
Cretaceous, Argentina
NHMUK V31414
3 cm across

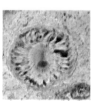

p.117
Confuciusornis sanctus
Lower Cretaceous, Liaoning
    Province, China
GMC V2131
Length from tuft on top of head
to bottom of
block 237.5 mm

p.119
Palaeooctopus newboldi
Cretaceous, Lebanon
NHMUK C32324
14 cm across

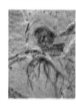

p.120
Coccoliths sp.
Upper Cretaceous, Sussex, UK
NHMUK (no registered number)
10 μm across

p.123
Polyblastidium racemosum
Upper Cretaceous, Oberg,
  Germany
NHMUK S2467
8 cm long

p.124
Tylocidaris clavigera
Upper Cretaceous, England, UK
NHMUK 39998
8 cm across

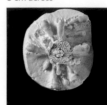

pp.126–127
Macropoma sp.
Upper Cretaceous, England, UK
NHMUK P4219
56 cm long

p.129
Troodon sp.
Upper Cretaceous, Montana, USA
USNM PAL358557
34 cm

pp.130–131
Edmontosaurus regalis
Upper Cretaceous, Red Deer
  River Valley, Alberta, Canada
NHMUK R8927
4 m long

p.132
**Poaceae (grass family),
with potential affinity to
subfamily Ehrhartoideae**
Upper Cretaceous, Pisdura,
  central India
BSIP 13162
103 μm long

pp.134–135
Mosasaurus sp.
Upper Cretaceous, Maastricht,
  Netherlands
NHMUK R1224
Cast 90 cm long

p.136
Torreites sanchezi
Upper Cretaceous, eastern central
  Oman
NHMUK LL28004
10 cm across

pp.138–139
Tyrannosaurus rex
Cretaceous, Cheyenne River,
    Weston Country, Wyoming,
    USA
NHMUK R8026
83 x 70 x 390 mm

pp.140–141
Belemnitella mucronata
Upper Cretaceous, Maastricht,
    Netherlands
NHMUK C8168
15 cm long

p.142
Radotruncana calcarata
Upper Cretaceous, Southeast
    Tanzania
USNM CN02L-18e
600 µm across

p.146
Otodus obliquus
Eocene, Oved Zem, Morocco
NHMUK P73114
35 cm across

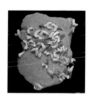

p.149
Onychonycteris finneyi
Eocene, Thompson Ranch
    Quarry, Wyoming, USA
AMNH FM-142467
Forearm length about 45 mm

pp.150–151
Rodhocetus sp.
Eocene, Pakistan
USNM PAL542461
37 cm

p.153
Leaf: *Byttnertiopsis*
daphnogenes
Fungal parasitoid:
Ophiocordyceps sp.
Eocene, Messel, Germany
SM.B.Me 10167
11 cm from preserved
    top of leaf to base
    of petiole

pp.154–155
Chaceon peruvianus
'Tertiary', Monte León,
    Argentina
NHMUK In28002
21 cm across

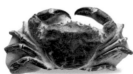

p.156
Nummulites gizehensis
Eocene, Giza, Egypt
NHMUK P36229
2.9 cm across

pp.158–159
Calyptogena sp.
Eocene, Spa, Barbados
LL31207
9.5 cm across

p.161
Trophon sowerbyi
Tertiary, Port San Julián,
Argentina
NHMUK G26415
5 cm across

pp.162–163
Basilosaurus cetoides
Eocene, Alabama, USA
USNM V13681
16.8 m

p.165
Margattea germari
Oligocene, 'East Prussia'
NHMUK I14476
Fossil 13 mm across

p.166
Aegyptopithecus zeuxis
Oligocene, Fayum, Egypt
NHMUK M36564
Model *c*. 10 cm long

p.169
Chaneya oeningensis
Miocene, Oeningen, Germany
NHMUK V17470
Fossil 2 cm across

p.170
Acer otopteryx
Miocene, Oeningen, Germany
NHMUK V52887
Fossil 9 cm across

p.172
Stylemys nebrascensis
Oligocene, White River
    Formation, Nebraska, USA
205 x 170 x 90 mm

p.175
Metrarabdotos thomseni
Plio-Pleistocene, Rhodes, Greece
NHMUK BZ4838
1 mm across

p.176
Proconsul africanus
Miocene, Rusinga Island, Kenya
NHMUK M51648 (cast)
Height 12 cm

pp.178–179
Patagornis marshi
Miocene, Santa Cruz Beds,
  Patagonia, Argentina
NHMUK A713
590 x 220 x 260 mm

p.180
Thylacosmilus atrox
Upper Miocene, Corral Quemado,
  Catamarca, Argentina
Field Museum, Chicago P14531
25 cm across

p.183
Discoglossus troscheli
Oligocene, Rott, Germany
NHMUK 35657
8 cm across

p.184
Shark coprolite
'Tertiary', East Anglia, UK
NHMUK Z696
10 cm high

p.187
***Cyprideis* sp.**
Middle Miocene, Şomuz
  Formation, Romania
Approximately 0.8 mm tall

p.189
Balanus concavus
Pliocene, Ramsholt, Suffolk, UK
NHMUK I763
6.5 cm across

p.190
Neptunea angulata
Plio-Pleistocene, Ramsholt,
  Suffolk, UK
NHMUK 74105
15 cm high

pp.192–193
Diprotodon bennetti
Pleistocene, Queensland,
  Australia
NHMUK M47855
40 cm across

pp.194–195
Glyptodon clavipes
Pleistocene, La Plata, Argentina
NHMUK M4473
2.5 m long

p.197
Australopithecus africanus
Plio-Pleistocene, Sterkfontein,
  South Africa
NHMUK EM3436 (cast)
Cast 12 cm across

p.198
The Laetoli footprints
Pliocene, Laetoli, Tanzania
Trackway 24 m long

pp.200–201
Myotragus balearicus
Pleistocene, Minorca, Spain
NHM M10962
15 cm across

pp.202–203
Megaloceros giganteus
Pleistocene, Bog of Ballybeta,
  Ireland
NHMUK M17959
3.2 m across

pp.204–205
Mammuthus trogontherii
Pleistocene, Ilford, UK
NHMUK M26543
2.5 m long

p.207
Dinornis sp.
Pleistocene, New Zealand
NHMUK A9039
37 cm across

p.209
Acropora cervicornis
Mid-Holocene, Bocas del Toro,
  Panama
AT-12-2-5
5.5 cm tall

pp.210–211
Hydrodamalis gigas
Holocene, Bering Sea
NHMUK (no registered number)
5 m long

p.212
Homo rhodesiensis
Pleistocene, Kabwe, Zambia
NHMUK E686
15 cm across

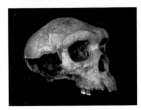

AMNH: American Museum of Natural History; NHMUK: Natural History Museum, London; RCCBYU: Yunnan University;
USNM: Smithsonian Institution, National Museum of Natural History; YPM: Yale Peabody Museum.
Reconstructions are included for some specimens to help appreciate what the living animal or plant may have looked like.

EON	ERA	PERIOD OR EPOCH	AGE (millions of years)
PHANEROZOIC	CENOZOIC	Holocene	0.012
		Pleistocene	2.6
		Pliocene	5.3
		Miocene	23
		Oligocene	34
		Eocene	56
		Paleocene	66
	MESOZOIC	Cretaceous	145
		Jurassic	201
		Triassic	252
	PALAEOZOIC	Permian	299
		Carboniferous	359
		Devonian	419
		Silurian	444
		Ordovician	485
		Cambrian	539
PRECAMBRIAN		Ediacaran	635
		Cryogenian	720
			4600

1. The vertical (time) axis is not to scale
2. Only the two youngest periods and none of the eras of the Precambrian are shown
3. Epochs rather than periods are specified for the Cenozoic

Index

Further information

BOOKS

Barthel, K.W., Swinburne, N.H.M. & Conway Morris, S., *Solnhofen. A Study in Mesozoic Palaeontology*. Cambridge University Press, Cambridge, 1990.

Benton, M.J., *Vertebrate Palaeontology*, 3rd edn. Wiley-Blackwell, Chichester, 2004.

Benton, M.J. & Harper, D.A.T. *Introduction to Palaeobiology and the Fossil Record*. Wiley-Blackwell, Chichester, 2020.

Briggs, D.E.G., Erwin, D.H. & Collier, F.J., *The Fossils of the Burgess Shale*. Smithsonian Institution Press, Washington DC, 1994.

Brusatte, S.L., *Dinosaur Paleobiology*. Wiley-Blackwell, Chichester, 2012.

Cowen, R., *History of Life,* 5th edn. Wiley-Blackwell, Chichester, 2013.

Dawkins, R., *The Ancestor's Tale*. Weidenfeld & Nicolson, London, 2004.

Fortey, R.A., *Life: An Unauthorised Biography*. HarperCollins, London, 1997.

Hou Xian-Guang, Aldridge, R.J., Bergström, J., Siveter, D.J., Siveter, D.J. & Feng Xiang-Hong., *The Cambrian Fossils of Chengjiang, China*. Blackwell, Oxford, 2004.

Humphrey, L. & Stringer, C., *Our Human Story*. Natural History Museum, London, 2022.

Kenrick, P., *A History of Plants in 50 Fossils*. Natural History Museum, London, 2020.

Knoll, A.H., *Life on a Young Planet*. Princeton University Press, Princeton, 2003.

Maisey, J.G., *Discovering Fossil Fishes*. Henry Holt and Company, New York, 1996.

Monks, N. & Palmer, C.P., *Ammonites*. Natural History Museum, London, 2003.

Naish, D., *Ancient Sea Reptiles*. Natural History Museum, London, 2023.

Naish, D. & Barrett, P.M., *Dinosaurs: How they lived and evolved*. Natural History Museum, London, 2023.

Prothero, D.R., *Evolution: What the Fossils Say and Why It Matters*. Columbia University Press, New York, 2007.

Ross, A., *Amber: The Natural Time Capsule*. Natural History Museum, London, 1998.

Ruse, M. & Travis, J. (eds.), *Evolution. The First Four Billion Years*. Belknap Press, Cambridge, 2009.

Savage, R.J.G. & Long, M.R., *Mammal Evolution*. Facts On File Publications, New York, 1986.

Schopf, J.W., *Cradle of Life*. Princeton University Press, Princeton, 1999.

Taylor, P.D. & Lewis, D.N., *Fossil Invertebrates*. Natural History Museum, London, 2005.

WEBSITES

Please note: all website addresses are subject to change.

NATURAL HISTORY MUSEUM DINOSAUR DIRECTORY:
https://www.nhm.ac.uk/discover/dino-directory.html

SMITHSONIAN NATIONAL MUSEUM OF NATURAL HISTORY, THE DEPARTMENT OF PALEOBIOLOGY
https://naturalhistory.si.edu/research/paleobiology

WIDE-RANGING US WEBSITE ON FOSSILS:
http://www.paleoportal.org

WEBSITE FOCUSING ON TRILOBITES:
http://www.trilobites.info

3D IMAGES OF BRITISH FOSSILS:
http://www.3d-fossils.ac.uk

Picture credits

pp. 7, 17, 73, 145 ©Ron Blakey, Colorado Plateau Geosystems, Inc.; pp. 8 top, 11, 20, 23, 25, 26, 28–9, 31, 33, 35, 37, 38, 40, 42, 44–5, 46–7, 49, 50–1, 52, 54–5, 57, 58, 61, 62, 68, 70–1, 75, 76, 79, 80–1, 83, 84–5, 86, 89, 90–1, 92–3, 94–5, 96, 99, 101, 102, 105, 106, 110, 114, 119, 120, 123, 124, 126–7, 130–1, 134–5, 136, 138–9, 140–1, 146, 154–5, 156, 158–9, 161, 165, 166, 169, 170, 172, 175, 176, 179, 183, 184, 189, 190, 192–3, 194–5, 197, 200–1, 202–3, 204–5, 207, 209, 210–11, 212, 214 (p. 11) stromatolite, 216 (p. 57) conulariid, 217 (pp. 80–1) *Glevosaurus* reconstructions, 218 (p. 89) living nautiloid, 218 (pp. 94–5) *Muraenosaurus*, 219 (p. 110) *Iguanodon*, and 224 (p. 207) *Dinornis* reconstructions ©NHM London; p. 8 bottom ©Bill Schopf, University of California Los Angeles; p. 12 ©John Cunningham, Bristol University; p. 15 ©Jim Gehling/South Australia Museum; p. 19 Alison Daley/National Type Collection of Invertebrate and Plant Fossils at the Geological Survey of Canada; pp. 65, 66–7, 129, 162–3 ©Donald E. Hurlberd, Smithsonian Institution, National Museum of Natural History; p. 109 courtesy of the Peabody Museum of Natural History, Yale University; p. 113 ©Enrique Peñalver, Instituto Geológico y Minero de España; p. 117 © The Geological Museum of China/NHM London; p. 132 © Caroline A. E. Strömberg; p. 142 ©Brian Huber, Smithsonian Institution, National Museum of Natural History; p. 149 ©AMNH; pp. 150–1 ©James Di Loreto, Smithsonian Institution, National Museum of Natural History; p. 153 ©Torsten Wappler, Steinmann Institut für Geologie, Mineralogie und Paläontologie, Universität Bonn; p. 180 courtesy of Ken Angielczyk, Field Museum of Natural History, Chicago; p. 187 ©Sergiu Loghin; p. 198 ©John Reader/Science Photo Library; pp. 214 (p. 19) *Anomalocaris*, 214 (p. 20) *Hallucigenia*, 216 (p. 52) *Pterichthys milleri*, 219 (p. 102) *Archaeopteryx*, and 223 (pp. 178–9) *Patagornis marshi* reconstructions ©John Sibbick/NHM London; p. 215 (pp. 28–9) *Orthoceras* courtesy of BBC; p. 215 (p. 33) conodont reconstruction ©Mark A. Purnell, University of Leicester; p. 216 (p. 58) *Lepidodendron* reconstruction © Philip Rye/NHM London; p. 216 (p. 61) *Glossopteris* reconstruction ©Rose Prevec/Science in Africa; pp. 216 (p. 62) *Eryops* reconstruction, 223 (pp. 192–3) *Diprotodon bennetti* reconstruction ©De Agostini/NHM London; pp. 217 (p. 65) *Helicoprion* reconstruction ©Florilegius/NHM London; p. 217 (pp. 76–7) *Massetognathus*, 217 (p. 79) *Megazostrodon rudnerae*, and 223 (p.180) *Thylacosmilus atrox* reconstructions ©Michael R. Long/NHM London; p. 218 (p. 99) *Leedsichthys problematicus* reconstruction ©Sciepro/Science Photo Library; p. 220 (pp. 134–5) *Mosasaurus* reconstruction ©Benislav Krzic/NHM London; p. 221 (pp. 140–1) belemnite reconstruction ©Richard Bizley/Science Photo Library.

NHM London ©The Trustees of the Natural History Museum, London.

All Rights Reserved; Picture Library www.piclib.nhm.ac.uk

Every effort has been made to contact and accurately credit all copyright holders. If we have been unsuccessful, we apologise and welcome corrections for future editions.

FURTHER INFORMATION

Acknowledgements

The authors would like to thank the following for reading parts of the text and for helping in the sourcing and imaging of specimens: Orangel Aguilera, Marcos Alvarez, Kenneth Angielczyk, Jane Barnbrook, Lisa Barnett, Eldredge Bermingham, Emma Bernard, Mary-Anne Binnie, David Bohaska, Trudy Brannan, Aude Caromel, Sandra Chapman, Alan Cheetham, Jennifer Clark, Anthony Coates, Richard Cooke, Simon Coppard, Allison Daley, Jill Darrell, Phil Donoghue, Greg Edgecombe, Doug Erwin, Tim Ewin, Howard Falcon-Lang, Seth Finnegan, Bevan M. French, Jim Gehling, Ian Glasspool, Walton Green, Ethan Grossman, Paul Harnik, Peta Hayes, Victoria Herridge, Ron Herzig, Jerry Hooker, Brian Huber, Zoe Hughes, Gene Hunt, Jeremy Jackson, Zerina Johanson, Kirk Johnson, Tom Jorstad, Paul Kenrick, Robert Krusynski, Damond Kyllo, Conrad Labandeira, Egbert Leigh, David Lindberg, Adrian Lister, S. Kathleen Lyons, Xiaoya Ma, Claire Mellish, Giles Miller, Matthew Miller, Scott Miller, Angela Milner, Noel Morris, Richard Norris, Kathleen O'Dea, Chris Page, Rachel Page, Lynne R. Parenti, Hugh Pearson, Richard Potts, Vandana Prasad, Oscar Puebla, Nicholas Pyenson, Gregory Retallack, Felix Rodriguez, Ira Rubinoff, Osmila Sanchez, JoAnn Sanner, Daniela Schmidt, Bill Schopf, Consuelo Sendino, Alessandra Serri, Nancy B. Simmons, William F. Simpson, Andrew Smith, Victor G. Springer, Lorna Steel, Chris Stringer, Caroline A. E. Strömberg, Hans-Dieter Sues, Elizabeth Helene Sweeny, Leif Tapanila, Harry Taylor (who photographed the majority of the Natural History Museum specimens with consummate skill), Patricia Taylor, Ellen Thomas, Jon Todd, Richard Twitchett, James Tyler, Torsten Wappler, Bill Wcislo, John Whittaker, Volker Wilde, Peter Wilf, Don Wilson, Samuel Zamora.